Autodesk
AutoCAD Architecture 2018
Fundamentals

Elise Moss

SDC
Publications

SDC Publications
P.O. Box 1334
Mission, KS 66222
913-262-2664
www.SDCpublications.com
Publisher: Stephen Schroff

ISBN-13: 978-1-63057-116-0
ISBN-10: 1-63057-116-4

Printed and bound in the United States of America.

Preface

No textbook can cover all the features in any software application. This textbook is meant for beginning users who want to gain a familiarity with the tools and interface of AutoCAD Architecture before they start exploring on their own. By the end of the text, users should feel comfortable enough to create a standard model, and even know how to customize the interface for their own use. Knowledge of basic AutoCAD and its commands is helpful but not required. I do try to explain as I go, but for the sake of brevity I concentrate on the tools specific to and within AutoCAD Architecture.

The files used in this text are accessible from the Internet from the book's page on the publisher's website: www.SDCpublications.com. They are free and available to students and teachers alike.

We value customer input. Please contact us with any comments, questions, or concerns about this text.

Elise Moss
elise_moss@mossdesigns.com

Acknowledgements from Elise Moss

This book would not have been possible without the support of some key Autodesk employees.

The effort and support of the editorial and production staff of SDC Publications is gratefully acknowledged. I especially thank Stephen Schroff for his helpful suggestions regarding the format of this text.

Finally, truly infinite thanks to Ari for his encouragement and his faith.

- Elise Moss

About the Author

Elise Moss has worked for the past thirty years as a mechanical designer in Silicon Valley, primarily creating sheet metal designs. She has written articles for Autodesk's Toplines magazine, AUGI's PaperSpace, DigitalCAD.com and Tenlinks.com. She is President of Moss Designs, creating custom applications and designs for corporate clients. She has taught CAD classes at DeAnza College, Silicon Valley College, and for Autodesk resellers. She is currently teaching CAD at Laney College in Oakland. Autodesk has named her as a Faculty of Distinction for the curriculum she has developed for Autodesk products. She holds a baccalaureate degree in Mechanical Engineering from San Jose State.

She is married with three sons. Her older son, Benjamin, is an electrical engineer. Her middle son, Daniel, works with AutoCAD Architecture in the construction industry. His designs have been featured in architectural journals. Her youngest son, Isaiah, currently attends community college. Her husband, Ari, is retired from a distinguished career in software development.

Elise is a third-generation engineer. Her father, Robert Moss, was a metallurgical engineer in the aerospace industry. Her grandfather, Solomon Kupperman, was a civil engineer for the City of Chicago.

She can be contacted via email at elise_moss@mossdesigns.com.

More information about the author and her work can be found on her website at www.mossdesigns.com.

Autodesk
Certified Instructor

Other books by Elise Moss
Autodesk Revit Architecture 2018 Basics
AutoCAD 2018 Fundamentals

Table of Contents

Notes:

Lesson 1:
Desktop Features

AutoCAD Architecture (ACA) enlists object oriented process systems (OOPS). That means that ACA uses intelligent objects to create a building. This is similar to using blocks in AutoCAD. Objects in AutoCAD Architecture are blocks on steroids. They have intelligence already embedded into them. A wall is not just a collection of lines. It represents a real wall. It can be constrained, has thickness and material properties, and is automatically included in your building schedule.

AEC is an acronym for Architectural/Engineering/Construction.
BID is an acronym for Building Industrial Design.
BIM is an acronym for Building Information Modeling.
AIA is an acronym for the American Institute of Architects.
MEP is an acronym for Mechanical/Electrical/Plumbing.

The following table describes the key features of objects in AutoCAD Architecture:

Feature Type	Description
AEC Camera	Create perspective views from various camera angles. Create avi files.
AEC Profiles	Create AEC objects using polylines to build doors, windows, etc.
Anchors and Layouts	Define a spatial relationship between objects. Create a layout of anchors on a curve or a grid to set a pattern of anchored objects, such as doors or columns.
Annotation	Set up special arrows, leaders, bar scales.
Ceiling Grids	Create reflected ceiling plans with grid layouts.
Column Grids	Define rectangular and radial building grids with columns and bubbles.
Design Center	Customize your AEC block library.
Display System	Control views for each AEC object.
Doors and Windows	Create custom door and window styles or use standard objects provided with the software.
Elevations and Sections	An elevation is basically a section view of a floor plan.
Layer Manager	Create layer standards based on AIA CAD Standards. Create groups of layers. Manage layers intelligently using Layer Filters.
Masking Blocks	Store a mask using a polyline object and attach to AEC objects to hide graphics.
Model Explorer	View a model and manage the content easily. Attach names to mass elements to assist in design.
Multi-view blocks	Blocks have embedded defined views to allow you to easily change view.

Feature Type	Description
Railings	Create or apply different railing styles to a stair or along a defined path.
Roofs	Create and apply various roof styles.
Floorplate slices	Generate the perimeter geometry of a building.
Spaces and Boundaries	Spaces and boundaries can include floor thickness, room height, and wall thickness.
Stairs	Create and apply various stair types.
Tags and Schedules	Place tags on objects to generate schedules. Schedules will automatically update when tags are modified, added, or deleted.
Template Files	Use templates to save time. Create a template with standard layers, text styles, linetypes, dimension styles, etc.
Walls	Create wall styles to determine material composition. Define end caps to control opening and end conditions. Define wall interference conditions.

AutoCAD Architecture sits on top of AutoCAD. It is helpful for users to have a basic understanding of AutoCAD before moving to AutoCAD Architecture. Users should be familiar with the following:

- AutoCAD toolbars and ribbons
- Zoom and move around the screen
- Manage blocks
- *Draw* and *Modify* commands
- *Model* and *Paper space* (layout)
- Dimensioning and how to create/modify a dimension style

If you are not familiar with these topics, you can still move forward with learning AutoCAD Architecture, but you may find it helpful to have a good AutoCAD textbook as reference in case you get stuck.

The best way to use AutoCAD Architecture is to start your project with massing tools (outside-in design) or space planning tools (inside-out design) and continue through to construction documentation.

The AEC Project Process Model

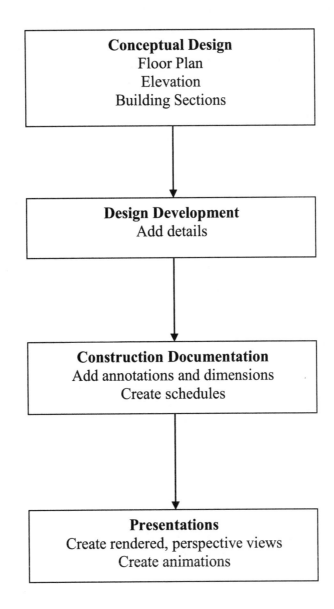

Conceptual Design

In the initial design phase, you can assemble AutoCAD Architecture mass elements as simple architectural shapes to form an exterior model of your building project. You can also lay out interior areas by arranging general spaces as you would in a bubble diagram. You can manipulate and consolidate three-dimensional mass elements into massing studies.

Later in this phase, you can create building footprints from the massing study by slicing floorplates, and you can begin defining the structure by converting space boundaries into walls. At the completion of the conceptual design phase, you have developed a workable schematic floor plan.

Design Development

As you refine the building project, you can add more detailed information to the schematic design. Use the features in AutoCAD Architecture to continue developing the design of the building project by organizing, defining, and assigning specific styles and attributes to building components.

Construction Development

After you have fully developed the building design, you can annotate your drawings with reference marks, notes, and dimensions. You can also add tags or labels associated with objects. Information from the objects and tags can be extracted, sorted, and compiled into schedules, reports, tables, and inventories for comprehensive and accurate construction documentation.

Presentations

A major part of any project is presenting it to the client. At this stage, you develop renderings, animations, and perspective views.

If you use the **QNEW** tool, it will automatically use the template set in the Options dialog.

Express Tools

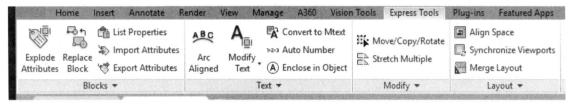

Express Tools are installed with AutoCAD Architecture. If you do not see Express Tools on the menu bar, try typing EXPRESSTOOLS to see if that loads them.

If for some reason you are missing the Express Tools you can modify the installation to add them.

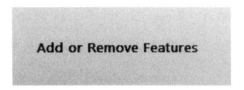

Go to the Control Panel, select **AutoCAD Architecture 2018** and select **Uninstall/Change**. Then select **Add or Remove Features**.

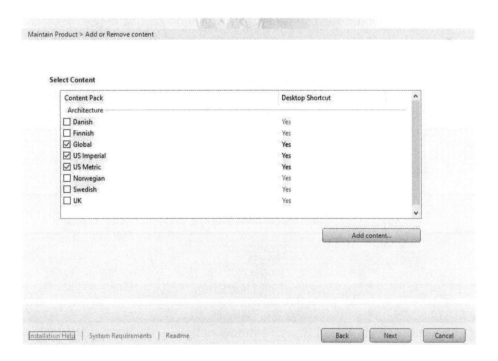

Press **Next** to skip this screen.

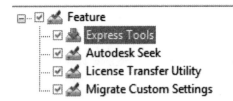

Place a check next to Express Tools to enable the installation and continue with the installation. Press **Update**.

Express Tools can be used to create custom line types, control layers, and many other useful functions.

If you forgot to install the Express Tools, they can be installed at any time. You don't need the Express Tools to run AutoCAD Architecture or to use this textbook. They just are useful for a lot of users.

Exercise 1-1:

Setting the Template for QNEW

Drawing Name: New
Estimated Time: 15 minutes

This exercise reinforces the following skills:

- ❑ Use of templates
- ❑ Getting the user familiar with tools and the ACA environment

1.   Launch ACA.

Select the drop-down arrow next to the large A to access the Application Menu.

Select **New→Drawing**.

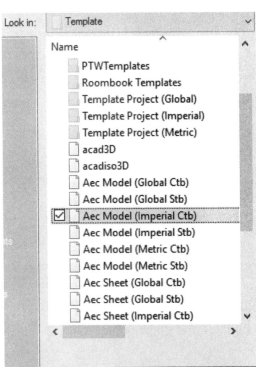

Select the desired template to use.

Each template controls units, layers, and plot settings.

Select *Aec Model (Imperial.Ctb).*

This template uses Imperial units (feet and inches and a Color Table to control plot settings).

Press **Open.**

2.   Place your cursor on the command line.

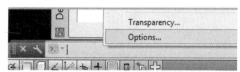   Right click the mouse.

Select **Options** from the short-cut menu.

3.   Select the **Files** tab.
Locate the *Template Settings* folder.

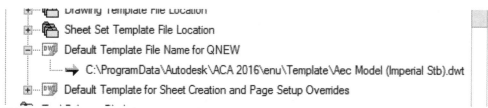

Click on the + symbol to expand.
Locate the Default Template File Name for QNEW.

4.   Browse...   Highlight the path and file name listed and select the **Browse** button.

5.   Browse to the *Template* folder.

This should be listed under

Program Data/Autodesk/ ACA 2018/enu

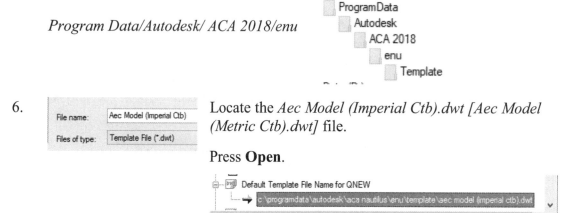

6.   Locate the *Aec Model (Imperial Ctb).dwt [Aec Model (Metric Ctb).dwt]* file.

Press **Open**.

7.   Note that you can also set where you can direct the user for templates. This is useful if you want to create standard templates and place them on a network server for all users to access.

You can also use this setting to help you locate where your template files are located.

8.   Press **Apply** and **OK**.

9.  Select the **QNEW** tool button.

10. 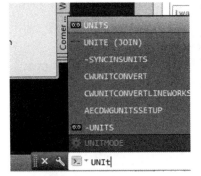 Note that you have tabs for all the open drawings.

    You can switch between any open drawing files by selecting the folder tab. You can also use the + tab to start a new drawing. The + tab acts the same way as the QNEW button and uses the same default template.

11. 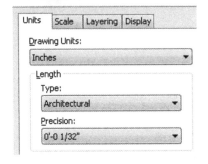 Type **units** on the *Command* line.

    Notice how you get a selection of commands as you type.

    Press ENTER to select UNITS.

12. Note that the units are in inches [millimeters].

    The units are determined by the template you set to start your new drawing.

13. Close the drawing without saving.

Templates can be used to preset layers, property set definitions, dimension styles, units, and layouts.

AEC Drawing Options

Menu	Tools→Options
Command line	Options
Context Menu→Options	Place mouse in the graphics area and right click
Shortcut	Place mouse in the command line and right click

Access the Options dialog box.

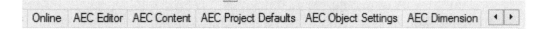

ACA's Options include five additional AEC specific tabs.
They are AEC Editor, AEC Content, AEC Project Defaults, AEC Object Settings, and AEC Dimension.

AEC Editor

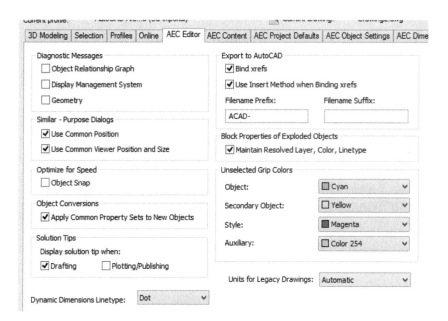

Diagnostic Messages	All diagnostic messages are turned off by default.
Similar-Purpose Dialogs	Options for the position of dialog boxes and viewers.
Use Common Position	Sets one common position on the screen for similar dialog boxes, such as door, wall, and window add or modify dialog boxes. Some dialog boxes, such as those for styles and properties, are always displayed in the center of the screen, regardless of this setting.
Use Common Viewer Position and Sizes	Sets one size and position on the screen for the similar-purpose viewers in AutoCAD Architecture. Viewer position is separately controlled for add, modify, style, and properties dialog boxes.
Optimize for Speed	Options for display representations and layer loading in the Layer Manager dialog.
Object Snap	Enable to limit certain display representations to respond only to the Node and Insert object snaps. This setting affects stair, railing, space boundary, multi-view block, masking block, slice, and clip volume result (building section) objects.
Apply Common Property Sets to New Objects	Property sets are used for the creation of schedule tables. Each AEC object has embedded property sets, such as width and height, to be used in its schedule. AEC properties are similar to attributes. Size is a common property across most AEC objects. Occasionally, you may use a tool on an existing object and the result may be that you replace the existing object with an entirely different object. For example, if you apply the tool properties of a door to an existing window, a new door object will replace the existing window. When enabled, any property sets that were assigned to the existing window will automatically be preserved and applied to the new door provided that the property set definitions make sense.
Solution Tips	Users can set whether they wish a solution tip to appear while they are drafting or plotting. A solution tip identifies a drafting error and suggests a solution.
Dynamic Dimensions Linetype	Set the linetype to be used when creating dynamic dimensions. This makes it easier to distinguish between dynamic and applied dimensions. You may select either continuous or dot linetypes.

Export to AutoCAD	Enable Bind Xrefs.
Export to AutoCAD ☑ Bind xrefs ☑ Use Insert Method when Binding xrefs Filename Prefix: Filename Suffix: ACAD-	If you enable this option, the xref will be inserted as a block and not an xref. Enable Use Insert Method when binding Xrefs if you want all objects from an xref drawing referenced in the file you export to be automatically exploded into the host drawing. If you enable this option, the drawing names of the xref drawings are discarded when the exported drawing is created. In addition, their layers and styles are incorporated into the host drawing. For example, all exploded walls, regardless of their source (host or xref) are located on the same layer. Disable Use Insert Method when binding Xrefs if you want to retain the xref identities, such as layer names, when you export a file to AutoCAD or to a DXF file. For example, the blocks that define walls in the host drawing are located on A-Wall in the exploded drawing. Walls in an attached xref drawing are located on a layer whose name is created from the drawing name and the layer name, such as Drawing1|WallA. Many architects automatically bind and explode their xrefs when sending drawings to customers to protect their intellectual property. Enter a prefix or a suffix to be added to the drawing filename when the drawing is exported to an AutoCAD drawing or a DXF file. In order for any of these options to apply, you have to use the Export to AutoCAD command. This is available under the Files menu.
Unselected Grip Colors	Assign the colors for each type of grip.

Units for Legacy Drawings	Determines the units to be used when opening an AutoCAD drawing in AutoCAD Architecture.
	    Automatic – uses the current AutoCAD Architecture units setting
	    Imperial – uses Imperial units
	    Metric – uses Metric units

Option Settings are applied to your current drawing and saved as the default settings for new drawings. Because AutoCAD Architecture operates in a Multiple Document Interface, each drawing stores the Options Settings used when it was created and last saved. Some users get confused because they open an existing drawing and it will not behave according to the current Options Settings.

AEC Content

| System | User Preferences | Drafting | 3D Modeling | Selection | Profiles | Online | AEC Editor | AEC Content |

Architectural / Documentation / Multi-Purpose Object Style Path:

C:\ProgramData\Autodesk\ACA 2018\enu\Styles\Imperial   [Browse...]

AEC DesignCenter Content Path:

C:\ProgramData\Autodesk\ACA 2018\enu\AEC Content   [Browse...]

Tool Catalog Content Root Path:

C:\ProgramData\Autodesk\ACA 2018\enu\   [Browse...]

IFC Content Path:

C:\ProgramData\Autodesk\ACA 2018\enu\Styles\   [Browse...]

Detail Component Databases: [Add/Remove...]

Keynote Databases: [Add/Remove...]

☑ Display Edit Property Data Dialog During Tag Insertion

Architectural/Documentation/Multipurpose Object Style Path	Allows you to specify a path of source files to be used to import object styles; usually used for system styles like walls, floors, or roofs.
AEC DesignCenter Content Path	Type the path and location of your content files or click Browse to search for the content files.
Tool Catalog Content Root Path	Type the path and location of your Tool Catalog files or click Browse to search for the content files.

IFC Content Path	Type the path and location of your Tool Catalog files or click Browse to search for the content files.
Detail Component Databases	Select the **Add/Remove** button to set the search paths for your detail component files.
Keynote Databases	Select the **Add/Remove** button to set the search paths for your keynote database files.
Display Edit Schedule Data Dialog During Tag Insertion	To attach schedule data to objects when you insert a schedule tag in the drawing, this should be ENABLED.

IFC Content is an object-based building data model that is non-proprietary. It uses open-source code developed by the IAI (International Alliance for Interoperability) to promote data exchange between CAD software. IFC Content might be created in Autodesk Revit, Inventor, or some other software for use in your projects.

Using AEC Content

AutoCAD Architecture uses several pre-defined and user-customizable content including:

- ❑ Architectural Display Configurations
- ❑ Architectural Profiles of Geometric Shapes
- ❑ Wall Styles and Endcap geometry
- ❑ Door styles
- ❑ Window styles
- ❑ Stair styles
- ❑ Space styles
- ❑ Schedule tables

Standard style content is stored in the AEC templates subdirectory. You can create additional content, import and export styles between drawings.

AEC Object Settings

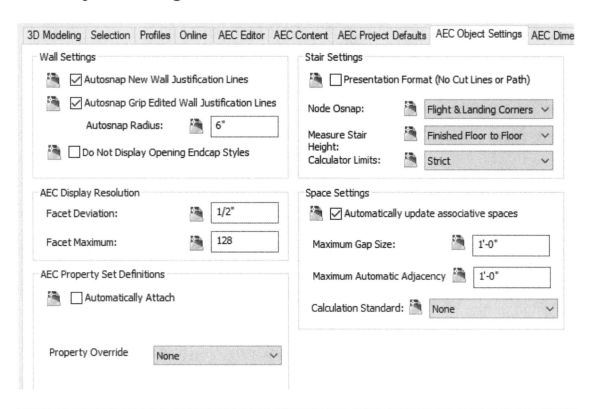

Wall Settings	
Autosnap New Wall Justification Lines	Enable to have the endpoint of a new wall that is drawn within the Autosnap Radius of the baseline of an existing wall automatically snap to that baseline. If you select this option and set your Autosnap Radius to 0, then only walls that touch clean up with each other.
Autosnap Grip Edited Wall Justification Lines	Enable to snap the endpoint of a wall that you grip edit within the Autosnap Radius of the baseline of an existing wall. If you select this option and set your Autosnap Radius to 0, then only walls that touch clean up with each other.
Autosnap Radius	Enter a value to set the snap tolerance.
Do not display opening end cap styles	Enable Do Not Display Opening Endcap Styles to suppress the display of endcaps applied to openings in walls. Enabling this option boosts drawing performance when the drawing contains many complex endcaps.

Stair Settings	Stair Settings ☐ Presentation Format (No Cut Lines or Path) Node Osnap: Flight & Landing Corners ⌄ Measure Stair Height: Finished Floor to Floor ⌄ Calculator Limits: Strict ⌄
Presentation Format (No Cut Lines or Path)	If this is enabled, a jagged line and directional arrows will not display.

Node Osnap	Determines which snap is enabled when creating stairs:
	    Vertical Alignment
	    Flight and Landing Corners

Measure Stair Height	Rough floor to floor – ignore offsets Finished floor to floor – include top and bottom offsets
Calculator Limits	Strict – stair will display a defect symbol when an edit results in a violation of the Calculation rules. Relaxed – no defect symbol will be displayed when the stair violates the Calculation rules. The Calculation rules determine how many treads are required based on the riser height and the overall height of the stairs.

AEC Display Resolution AEC Display Resolution Facet Deviation: 1/2" Facet Maximum: 128	This determines the resolution of arcs and circular elements. Facet Deviation – The default is ½″. Facet Maximum – this can be set from 100 to 10,000. Higher settings use more memory and may affect screen refresh rates.

Space Settings 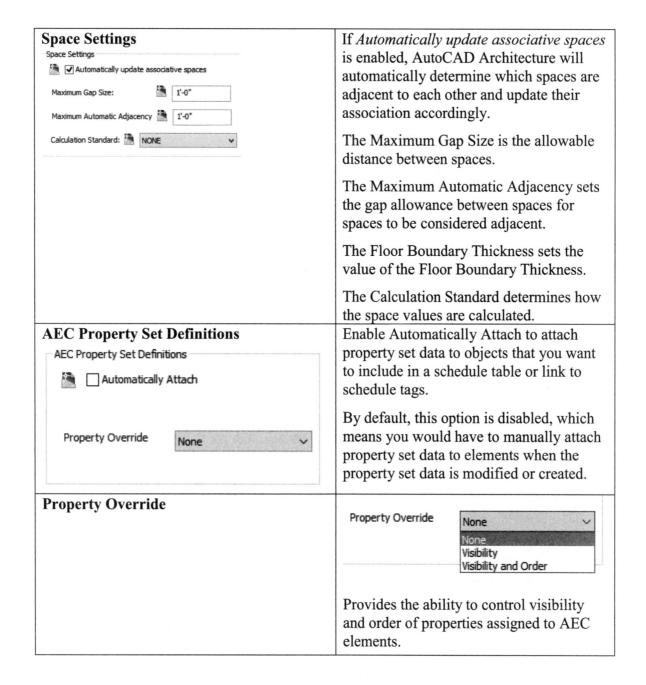	If *Automatically update associative spaces* is enabled, AutoCAD Architecture will automatically determine which spaces are adjacent to each other and update their association accordingly. The Maximum Gap Size is the allowable distance between spaces. The Maximum Automatic Adjacency sets the gap allowance between spaces for spaces to be considered adjacent. The Floor Boundary Thickness sets the value of the Floor Boundary Thickness. The Calculation Standard determines how the space values are calculated.
AEC Property Set Definitions	Enable Automatically Attach to attach property set data to objects that you want to include in a schedule table or link to schedule tags. By default, this option is disabled, which means you would have to manually attach property set data to elements when the property set data is modified or created.
Property Override	Provides the ability to control visibility and order of properties assigned to AEC elements.

AEC Dimension

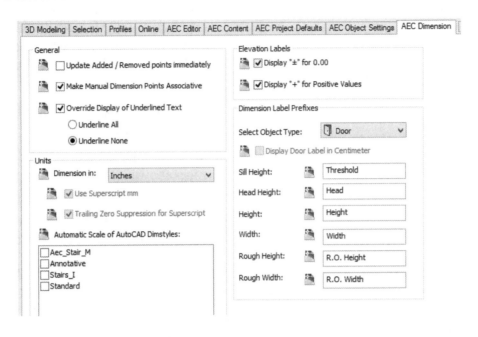

General	
Update Added/Removed points immediately	Enable to update the display every time you add or remove a point from a dimension chain.
Make Manual Dimension Points Associative	Enable if you want the dimension to update when the object is modified.
Override Display of Undefined Text	Enable to automatically underline each manually overridden dimension value.
Underline All	Enable to manually underline overridden dimension values.
Underline None	Enable to not underline any overridden dimension value.
Units	
Dimension in:	Select the desired units from the drop-down list.
Use Superscript mm	If your units are set to meters or centimeters, enable superscripted text to display the millimeters as superscript.

Trailing Zero Suppression for Superscript	To suppress zeros at the end of superscripted numbers, select Trailing Zeros Suppression for Superscript. You can select this option only if you have selected Use Superscript mm and your units are metric. 4.12^3
Automatic Scale of AutoCAD Dimstyles	Select the dimstyles you would like to automatically scale when units are reset.
Elevation Labels	Select the unit in which elevation labels are to be displayed. This unit can differ from the drawing unit.
Display +/- for 0.00 Values	
Display + for Positive Values	
Dimension Label Prefixes	Select Object Type: Door — Door Window Opening Stair Display Door Lab Sill Height: Dimension Label Prefixes are set based on the object selected from the drop-down. You can set label prefixes for Doors, Windows, Openings, and Stairs. Select the object, and then set the prefix for each designation. The designations will change based on the object selected.

AEC Project Defaults

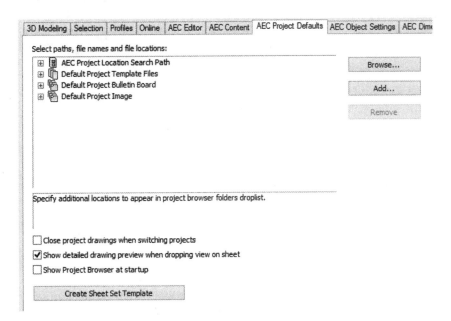

Most BID/AEC firms work in a team environment where they divide a project's tasks among several different users. The AEC Project Defaults allows the CAD Manager or team leader to select a folder on the company's server to locate files, set up template files with the project's title blocks, and even set up a webpage to post project information.

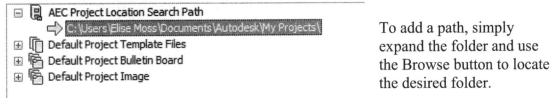

To add a path, simply expand the folder and use the Browse button to locate the desired folder.

You can add more than one path to the AEC Project Location Search Path. All the other folders only allow you to have a single location for template files, the project bulletin board, and the default project image.

I highly recommend that you store any custom content in a separate directory as far away from AutoCAD Architecture as possible. This will allow you to back up your custom work easily and prevent accidental erasure if you upgrade your software application.

You should be aware that AutoCAD Architecture currently allows the user to specify only ONE path for custom content, so drawings with custom commands will only work if they reside in the specified path.

☐ Close project drawings when switching projects	If enabled, when a user switches projects in the Project Navigator, any open drawings not related to the new project selected will be closed. This conserves memory resources.
☑ Show detailed drawing preview when dropping view on sheet	If enabled, then the user will see the entire view as it is being positioned on a sheet. If it is disabled, the user will just see the boundary of the view for the purposes of placing it on a sheet.
☐ Show Project Browser at startup	If enabled, you will automatically have access to the Project Browser whenever you start AutoCAD Architecture.

Accessing the Project Browser

Command line	AECProjectBrowser
Quick Access Toolbar	
Files Menu	Project Browser

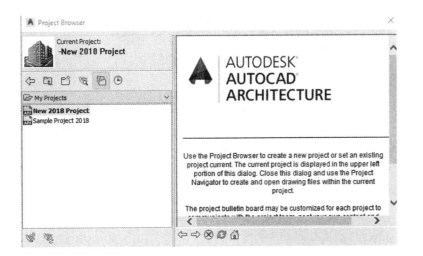

The Project Browser allows you to manage your project. Projects have two parts: the building model and the reports generated from the building model.

The building model is made of two drawing types: constructs and elements. A construct is any unique portion of a building. It can be a flight of stairs, a specific room, or an entire floor. Constructs are assigned to a level (floor) and a division within a project. Elements are any AEC object or content that is used multiple times in a model. For example, furniture layouts or lavatory layouts. Elements can be converted into constructs if you decide that they will not be used repeatedly.

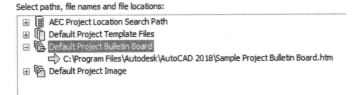

The text you see in the right window is actually an html file. The path to this file is controlled in the path for the Default Project Bulletin Board. This is set on the AEC Project Defaults tab in the Options dialog.

Architectural Profiles of Geometric Shapes

You can create mass elements to define the shape and configuration of your preliminary study, or mass model. After you create the mass elements you need, you can change their size as necessary to reflect the building design.

- **Mass element**: A single object that has behaviors based on its shape. For example, you can set the width, depth, and height of a box mass element, and the radius and height of a cylinder mass element.

Mass elements are parametric, which allows each of the shapes to have very specific behavior when it comes to the manipulation of each mass element's shape. For example, if the corner grip point of a box is selected and dragged, then the width and depth are modified. It is easy to change the shape to another form by right-clicking on the element and selecting a new shape from the list.

Through Boolean operations (addition, subtraction, intersection), mass elements can be combined into a mass group. The mass group provides a representation of your building during the concept phase of your project.

- **Mass group**: Takes the shape of the mass elements and is placed on a separate layer from the mass elements.
- **Mass model**: A virtual mass object, shaped from mass elements, which defines the basic structure and proportion of your building. A marker appears as a small box in your drawing to which you attach mass elements.

As you continue developing your mass model, you can combine mass elements into mass groups and create complex building shapes through addition, subtraction, or intersection of mass elements. You can still edit individual mass elements attached to a mass group to further refine the building model.

To study alternative design schemes, you can create a number of mass element references. When you change the original of the referenced mass element, all the instances of the mass element references are updated.

The mass model that you create with mass elements and mass groups is a refinement of your original idea that you carry forward into the next phase of the project, in which you change the mass study into floor plates and then into walls. The walls are used to start the design phase.

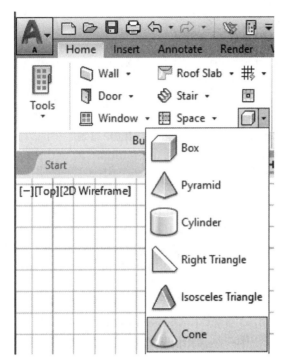

Mass Elements are accessed from the Home ribbon.

Mass Elements that can be defined from the ribbon include Arches, Gables, Doric Columns, etc.

The Style Manager

Access the Style Manager from the Manage ribbon.

The Style Manager is a Microsoft® Windows Explorer-based utility that provides you with a central location in Autodesk AutoCAD Architecture where you can view and work with styles in drawings or from Internet and intranet sites.

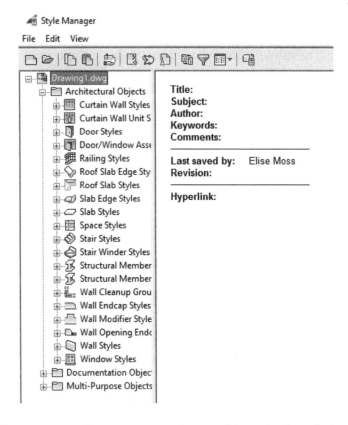

Styles are sets of parameters that you can assign to objects in Autodesk AutoCAD Architecture to determine their appearance or function. For example, a door style in Autodesk AutoCAD Architecture determines what door type, such as single or double, bi-fold or hinged, a door in a drawing represents. You can assign one style to more than one object, and you can modify the style to change all the objects that are assigned that style.

Depending on your design projects, either you or your CAD Manager might want to customize existing styles or create new styles. The Style Manager allows you to easily create, customize, and share styles with other users. With the Style Manager, you can:

- Provide a central point for accessing styles from open drawings and Internet and intranet sites

- Quickly set up new drawings and templates by copying styles from other drawings or templates

- Sort and view the styles in your drawings and templates by drawing or by style type

- Preview an object with a selected style

- Create new styles and edit existing styles

- Delete unused styles from drawings and templates

- Send styles to other Autodesk AutoCAD Architecture users by email

Objects in Autodesk AutoCAD Architecture that use styles include 2D sections and elevations, AEC polygons, curtain walls, curtain wall units, doors, endcaps, railings, roof slab edges, roof slabs, schedule tables, slab edges, slabs, spaces, stairs, structural members, wall modifiers, walls, window assemblies, and windows.

Additionally, layer key styles, schedule data formats, and cleanup group, mask block, multi-view block, profile, and property set definitions are handled by the Style Manager.

Most of the objects in Autodesk AutoCAD Architecture have a default Standard style. In addition, Autodesk AutoCAD Architecture includes a starter set of styles that you can use with your drawings. The Autodesk AutoCAD Architecture templates contain some of these styles. Any drawing that you start from one of the templates includes these styles.

You can also access the styles for doors, end caps, spaces, stairs, walls, and windows from drawings located in *ProgramData\Autodesk\ACA 2018\enu\Styles*.

.Property set definitions and schedule tables are located in the *ProgramData\Autodesk\ACA 2018\enu\AEC Content* folder.

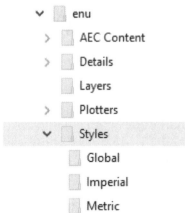

Exercise 1-2:
Creating a New Geometric Profile

Drawing Name: New
Estimated Time: 15 minutes

This exercise reinforces the following skills:

□   Use of AEC Design Content
□   Use of Mass Elements
□   Use of Views
□   Visual Styles
□   Modify using Properties

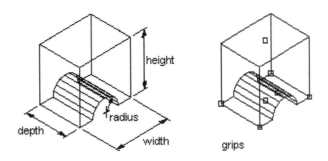

Creating an arch mass element

1.   Start a New Drawing.

2.   Select the QNEW tool from the Standard toolbar.

3.   Because we assigned a template in Exercise 1-1, a drawing file opens without prompting us to select a template.

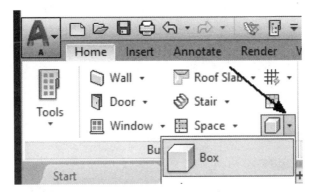

All the massing tools are available on the **Home** ribbon in the Build section.

4.   Select the **Arch** tool from the Massing drop down list.

5.   Expand the **Dimensions** section on the Properties palette.

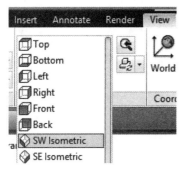

Set Specify on screen to **No**.
Set the Width to **2′ [600]**.
Set the Depth to **6″ [200]**.
Set the Height to **1′ 6″ [500]**.
Set the Radius to **6″ [200]**.

6.

Pick a point anywhere in the drawing area.

Press ENTER to accept a Rotation Angle of 0 or right click and select **ENTER**.

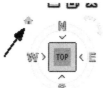

Press **ENTER** to exit the command.

7. Switch to an isometric view.

To switch to an isometric view, simply go to **View** Ribbon. Under the Views list, select **SW Isometric**.

Or select the Home icon next to the Viewcube.

Our model so far.

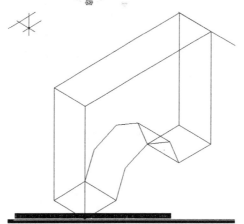

8.

Select **Realistic** on the View ribbon under Face Effects to shade the model.

You can also type **SHA** on the command line, then **R** for Realistic.

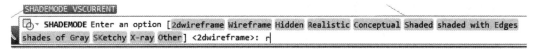

9. To change the arch properties, select the arch so that it is highlighted. If the properties dialog doesn't pop up automatically, right click and select **Properties**.

10.

Dimensions	
Width	2'-0"
Depth	6"
Height	1'-6"
Radius	9"
Volume	1 CF

Change the Radius to **9″** [**250**].

Pick into the graphics window and the arch will update.

11. Press **ESC** to release the grips on the arch.

12. Save the drawing as *ex1-2.dwg*.

Save Drawing - Version Conflict

This drawing contains custom objects that are not supported in previous versions. These objects cannot be saved to a previous version. You can do one of the following things.

Save this drawing to the current version:
Objects will become incompatible with earlier versions.

Save this drawing as AutoCAD-only objects:
To view AutoCAD objects in a previous version verify that the latest object enabler is installed. (Available from www.autodesk.com\enablers)

Save this drawing to previous version with Proxy Graphics:
If you open this drawing in a previous version you can view AEC objects as Proxy Objects but cannot edit them.

☑ Do not show me this message again   [Close]

Click here for more information on Sharing Drawings with AutoCAD Users

If you see this dialog, it means that the mass element just created (the arch) cannot be saved as a previous version file.

Press **Close**.

Visual Styles are stored with the drawing.

Exercise 1-3:

Creating a New Visual Style

Drawing Name: ex1-2.dwg
Estimated Time: 15 minutes

This exercise reinforces the following skills:

- ❑ Workspaces
- ❑ Use of Visual Styles
- ❑ Controlling the Display of Objects

1. Open *ex1-2.dwg*.

2.  Activate the **View** ribbon.

3.  Launch the Visual Styles Manager.

   To launch, select the small arrow in the lower right corner of the Visual Styles panel.

4. 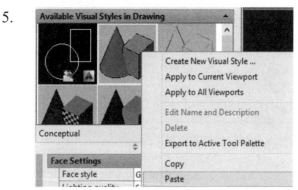 Highlight the **Conceptual** tool.

   Right click and select **Copy**.

5. Right click and select **Paste**.

6.  Highlight the copied tool.
Right click and select **Edit Name and Description**.

7. Edit Name and Description

Name: Real Conceptual

Description: Real Conceptual

OK

Change the name and description to **Real Conceptual**.

Press **OK**.

8.
Face Settings	
Face style	Gooch
Lighting quality	Smooth
Color	Normal
Monochrome color	255,255,255
Opacity	-60
Material display	Materials and textures

Under Face Settings:
Set Material display to **Materials and textures**.

9.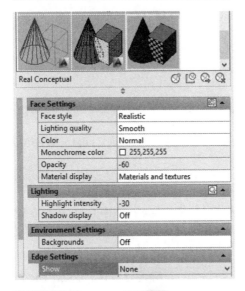

Real Conceptual

Face Settings	
Face style	Realistic
Lighting quality	Smooth
Color	Normal
Monochrome color	255,255,255
Opacity	-60
Material display	Materials and textures

Lighting	
Highlight intensity	-30
Shadow display	Off

Environment Settings	
Backgrounds	Off

Edge Settings	
Show	None

Under Face Settings:

Set the Face Style to **Realistic**.

Under Environment Settings:

Set the Backgrounds to **Off**.

Under Edge Settings:

Set the Show to **None**.

10. Create New Visual Style ...
Apply to Current Viewport
Apply to All Viewports

Verify that the Real Conceptual Style is highlighted/ selected.

Right click and select the **Apply to Current Viewport** tool.

[−][Custom View][Real Conceptual]

Note that the active visual style is listed in the upper right corner of the drawing.

11. Save the drawing as *ex1-3.dwg*.

Exercise 1-4:
Creating an Autodesk 360 Account

Drawing Name: (none, start from scratch)
Estimated Time: 5 Minutes

This exercise reinforces the following skills:

❏ Create an account on Autodesk 360.

You must have an internet connection to access the Autodesk server.

1.   Next to the Help icon at the top of the screen:

Select **Sign In to Autodesk account**.

2.   If you have an existing Autodesk ID, you can use it.

If not, select the link that says **Create Account.**

You may have an existing Autodesk ID if you have registered Autodesk software in the past. Autodesk 360 is free to students and educators. For other users, you are provided a small amount of initial storage space for free.

I like my students to have a 360 account because it will automatically back up their work.

3.

Create account

First name   Last name

Email

Confirm email

Password

☐ I agree to the A360 Terms of Service and the Autodesk Privacy Statement.

CREATE ACCOUNT

ALREADY HAVE AN ACCOUNT? SIGN IN

Fill in the form to create an Autodesk ID account.

Be sure to write down the ID you select and the password.

4.

Account created

This single account gives you access to all your Autodesk products

☐ I would like to receive email communications from Autodesk

DONE

You should see a confirmation window stating that you now have an account.

5.

Welcome, Elise Moss
AutoCAD Architecture 2018

Choose your program preferences for working with A360. You can also change these settings from the Online tab of the Options dialog box.
Learn more about A360

Custom Settings Sync

Sync your program appearance, profiles, workspaces, options, and support files, so you can work in a familiar environment on any computer that has this program installed.
Choose which settings are synced

☑ Sync my settings

OK   Help

Once the account is created, you will automatically be logged in to your account.

There will be a slight pause while the server validates the log in.

You will then be asked to check your settings.

You can automatically save your work to the Cloud.
If you do this, you will be able to access your files anywhere you have internet access.

Press **OK**.

6. You will see the name you selected as your Autodesk ID in the Cloud sign-in area.

Tips & Tricks

Use your Autodesk 360 account to create renderings, back-up files to the Cloud so you can access your drawings on any internet-connected device, and to share your drawings.

Exercise 1-5:

Creating a New Wall Style

Drawing Name: New using Imperial [Metric] (ctb).dwt
Estimated Time: 30 minutes

This exercise reinforces the following skills:

- ❑ Style Manager
- ❑ Use of Wall Styles

1.   Select the **QNEW** tool from the Standard toolbar.

2. 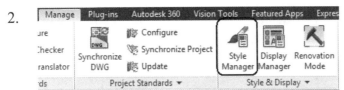   Activate the **Manage** ribbon.

    Select the **Style Manager** from the **Style & Display** panel.

3.   The Style Manager is displayed with the current drawing expanded in the tree view. The wall styles in the current drawing are displayed under the wall style type. All other style and definition types are filtered out in the tree view.

    Locate the wall styles in the active drawing.

4.   Highlight Wall Styles by left clicking on the name.
    Right click and select **New**.

5.   Change the name of the new style by typing **Brick_Block**.

6.   In the General tab, type in the description **8″ CMU and 3-1/2″ [200 mm CMU and 90 mm] Brick Wall** in the Description field as shown.

7. Select the Components tab.
    Change the Name to **CMU** by typing in the Name field indicated.

    Press the dropdown arrow next to
    the Edge Offset Button.
    Set the Edge Offset to 0.

    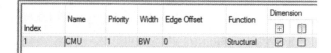

8. A 0 edge offset specifies that the outside edge of the CMU is coincident with the
    wall baseline.

9. Press the **Width** down arrow button.

    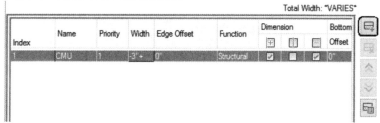

    Set the value to **-3″**
    [**-80**].

    This specifies that the CMU has a fixed width of 3 inches [80 mm] in the negative
    direction from the wall baseline (to the inside).

    You now have one component defined for your wall named CMU with the
    properties shown.

10. Press the **Add** Component button on the right side of the dialog box.
    Next, we'll add wall insulation.

    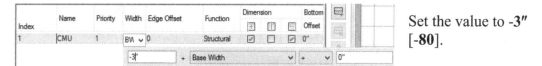

11. In the Name field, type **Insulation**.

    Set the Priority to 1. (The lower the priority number, the higher the priority when
    creating intersections.)

    Set the Edge Offset with a value of **0**.

    Set the Component Width as shown, with a value of **1.5″** [**38**] and the Base Width
    set to 0.

    This specifies that the insulation has a fixed width of 1.5″ [38 mm] offset in a
    positive direction from the wall baseline (to the outside).

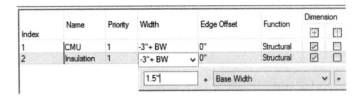

12. 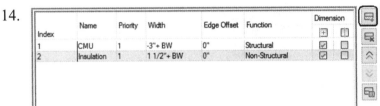 Set the Function to **Non-Structural**.

13.  Note that the preview changes to show the insulation component.

14.  Now, we add an air space component to our wall. Press the **Add Component** button.

15. Rename the new component **Air Space**.

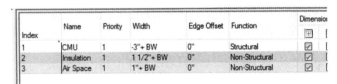 Set the Edge Offset to **0**.

This specifies that the inside edge of the air space is coincident with the outside edge of the insulation.

The width should be set to **1″ [25.4] + Base Width**.

This specifies that the air gap will have a fixed width of 1 inch [25.4 mm] offset in the positive direction from the wall baseline (to the outside).

Index	Name	Priority	Width	Edge Offset	Function	Dimension
1	CMU	1	-3″+ BW	0″	Structural	☑
2	Insulation	1	1 1/2″+ BW	0″	Non-Structural	☑
3	Air Space	1	1″+ BW	0″	Non-Structural	☑

16. Use the **Add Component** tool to add the fourth component.

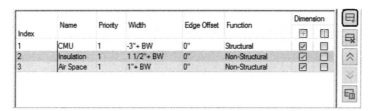

Index	Name	Priority	Width	Edge Offset	Function	Dimension	
						⊞	⬚
1	CMU	1	-3"+ BW	0"	Structural	☑	☐
2	Insulation	1	1 1/2"+ BW	0"	Non-Structural	☑	☐
3	Air Space	1	1"+ BW	0"	Non-Structural	☑	☐

The Name should be **Brick**.

The Priority set to **1**.

The Edge Offset set to **2.5″ [63.5]**.

The Width set to **3.5″ [90]**.

The Function should be set to **Structural**.

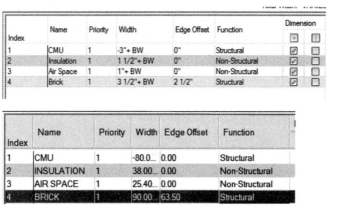

Index	Name	Priority	Width	Edge Offset	Function	Dimension	
						⊞	⬚
1	CMU	1	-3"+ BW	0"	Structural	☑	☐
2	Insulation	1	1 1/2"+ BW	0"	Non-Structural	☑	☐
3	Air Space	1	1"+ BW	0"	Non-Structural	☑	☐
4	Brick	1	3 1/2"+ BW	2 1/2"	Structural	☑	☐

Index	Name	Priority	Width	Edge Offset	Function	
1	CMU	1	-80.0...	0.00	Structural	
2	INSULATION	1	38.00...	0.00	Non-Structural	
3	AIR SPACE	1	25.40...	0.00	Non-Structural	
4	BRICK	1	90.00...	63.50	Structural	

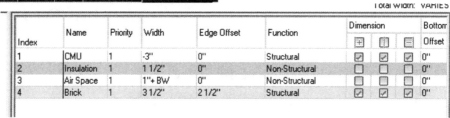

Total Width: VARIES

Index	Name	Priority	Width	Edge Offset	Function	Dimension			Bottom Offset
						⊞	⬚	⊟	
1	CMU	1	-3"	0"	Structural	☑	☑	☑	0"
2	Insulation	1	1 1/2"	0"	Non-Structural	☐	☐	☐	0"
3	Air Space	1	1"+ BW	0"	Non-Structural	☐	☐	☐	0"
4	Brick	1	3 1/2"	2 1/2"	Structural	☑	☑	☑	0"

17. The Dimension section allows you to define where the extension lines for your dimension will be placed when measuring the wall. In this case, we are not interested in including the air space or insulation in our linear dimensions. Uncheck those boxes.

18. 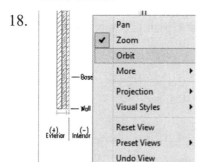 Note how your wall style previews.

If you right click in the preview pane, you can change the preview window by zooming, panning, or assigning a Preset View.

The preview pane uses a DWF-style interface.

19. 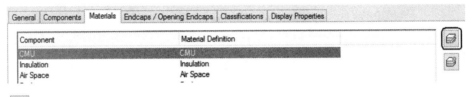 Select the **Materials** tab.

Highlight the **CMU** component.

Select the **Add New Material** tool.

20. 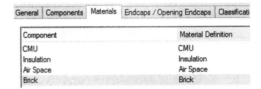 Enter **CMU** in the New Name field and press **OK**.

21. Use the **Add New Material** tool to assign new materials to each component.

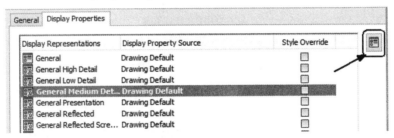

22. Highlight the **CMU** Component.

Component	Material Definition	
CMU	CMU	
Insulation	Insulation	
Air Space	Air Space	

General | Components | Materials | Endcaps / Opening Endcaps | Classifications | Display Properties

Select the **Edit Material** tool.

23. Select the **Display Properties** tab.

Display Representations	Display Property Source	Style Override	
General	Drawing Default	☐	
General High Detail	Drawing Default	☐	
General Low Detail	Drawing Default	☐	
General Medium Det...	Drawing Default	☐	
General Presentation	Drawing Default	☐	
General Reflected	Drawing Default	☐	
General Reflected Scre...	Drawing Default	☐	

General | Display Properties

Highlight the **General Medium Detail** Display Representation.

Click on the **Style Override** button.

24. Select the **Hatching** tab.

Highlight **Plan Hatch** and select **Pattern**.

Under Type, select **Predefined**.

Under Pattern Name, select **AR-CONC**.

25. Assign the **AR-CONC** Pattern to all the Display Components.

Set the Angles to **0**.

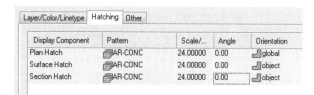

26. Press **OK** twice to return to the Materials tab of the Wall Properties dialog.

27.  Highlight **Insulation** and select the **Edit Material** tool.

28. Select the **Display Properties** tab.

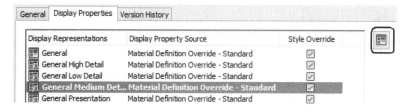

Click on the **Style Override** button.

Highlight the **General Medium Detail** Display Property.

29. Select the **Hatching** tab.

Select all the Display Components so they are all highlighted.

30.

Set the Plan Hatch to the **INSUL** pattern under Predefined type.

Set the Angle to **0.00**.

31.

Verify that a check is placed in the Style Override button.

32. Press **OK** to return to the previous dialog.

33. Highlight **Brick** and select the **Edit Material** tool.

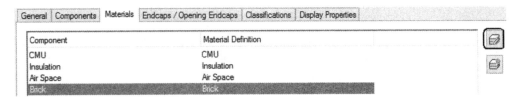

34. Select the **Display Properties** tab.

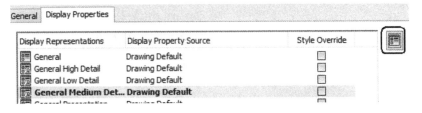

Highlight the **General Medium Detail** Display Representation.

Click on the **Style Override** button.

35. 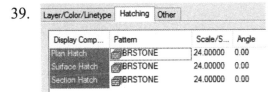 Select the **Hatching** tab.

Highlight the three display components.

Select the first pattern.

36. Set the Type to **Predefined**.

37. Select the **Browse** button.

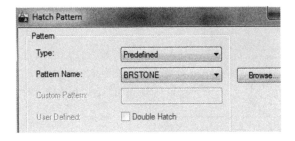

38. Select the **BRSTONE** pattern.

BRSTONE

39. Set the Angle to 0 for all hatches.

Display Comp...	Pattern	Scale/S...	Angle
Plan Hatch	BRSTONE	24.00000	0.00
Surface Hatch	BRSTONE	24.00000	0.00
Section Hatch	BRSTONE	24.00000	0.00

Press **OK** twice.

40.

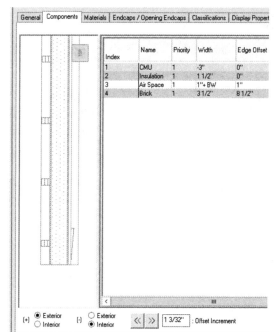

Activate the Components tab. Highlight each component and use the << >> arrows to position the component. By shifting each component position, you can see where it is located. Locate the Brick so it is on the Exterior side and the CMU so it is interior.

☑ Auto Calculate Edge Offset

You can also use the Auto Calculate Edge Offset option to ensure all the components are correct.

41. Close the Style Manager.

Press **OK** to exit the Edit dialog.
You can preview how the changes you made to the material display will appear when you create your wall.

42. Save as *ex1-4.dwg*.

Exercise 1-6:
Tool Palettes

Drawing Name: ex1-4.dwg
Estimated Time: 15 minutes

This exercise reinforces the following skills:

    ❑ Use of AEC Design Content
    ❑ Use of Wall Styles

1.   Open or continue working in *ex1-4.dwg*.

2.  Activate the Home ribbon.

    Select **Tools→Design Tools** to launch the tool palette.

3. 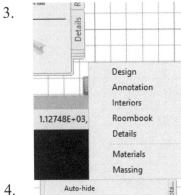 Right click on the tabs and you will see a list of the tabs available in the Design Tools.

4.  Right click on the title bar on the left side of the tool palette.

    Select **New Palette**.

5. Right click on the blank palette.

    Select **Rename Palett**e.

6. 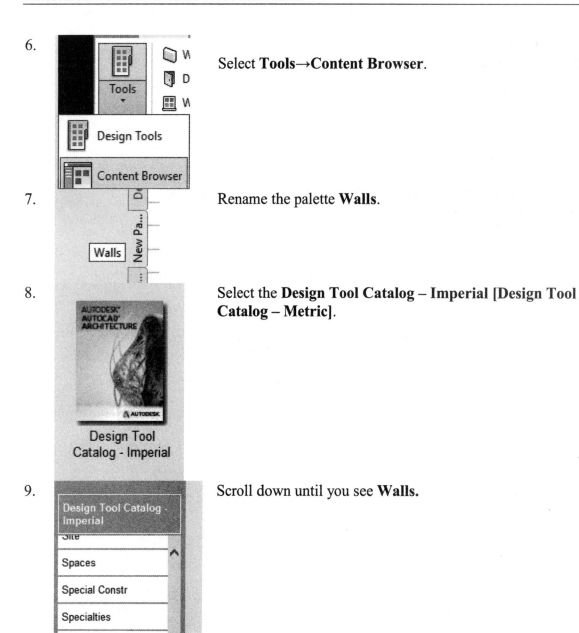 Select **Tools→Content Browser**.

7. Rename the palette **Walls**.

8. Select the **Design Tool Catalog – Imperial [Design Tool Catalog – Metric]**.

9. Scroll down until you see **Walls.**

10. 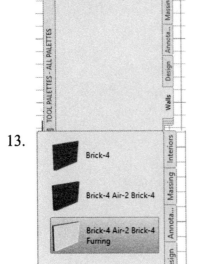 Click on the **Brick** link.

11. Put your mouse over the blue I next to the **Brick-4[Brick-090]** wall style.

You should see an eyedropper.

Hold down the left mouse button.
Drag over to the new Walls palette.
Release the left mouse button.

12. The wall style is now added to the Walls palette.

13. Repeat to add the **Brick 4-Air 2 Brick-4 [Brick-090-050 Brick-090]** and **Brick -4 Air- 2 Brick – 4 Furring [Brick-090 Air-050 Brick-090 Furring]** wall styles.

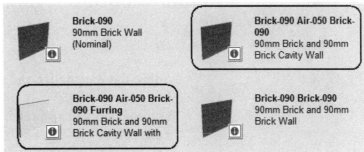

14. Select the back button.

15. Select the **Stud** folder.

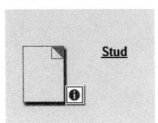

16.

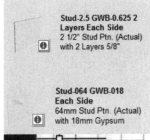

Drag and drop the **Stud -2.5 GWB-0.625 Each Side [Stud-064 GWB-018 Each Side]** wall style to the Walls palette.

17.

On the Walls tool palette:

Select the **Properties** button.

18.

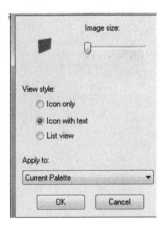

Select **View Options**.

19.

Adjust the Image Size and select the desired View style.

You can apply the preference to the Current Palette (the walls palette only) or all palettes.

Press **OK**.

20. 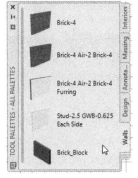 Verify that the tool palette is open with the Walls palette active.

21. Activate the **Manage** ribbon.

Select the **Style Manager** tool.

22. Locate the **Brick-Block** wall style you created in the Style Manager.

23. 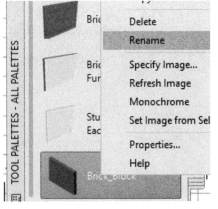 Drag and drop the wall style onto your Walls palette.

Press **OK** to close the Style Manager.

24. Highlight the **Brick_Block** tool.
Right click and select **Rename**.

25. Change the name to **8″ CMU- 3.5″ Brick [200 mm CMU – 90 mm]**.

26. Select the new tool and draw a wall.
You may draw a wall simply by picking two points.
If you enable the ORTHO button on the bottom of the screen, your wall will be straight.

27.
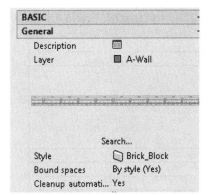
Select the wall you drew.

Right click and select **Properties**.

Note that the wall is placed on the A-Wall Layer and uses the Brick-Block Style you defined.

28. Save your file as *Styles.dwg*.

The wall style you created will not be available in any new drawings unless you copy it over to the default template.

 To make it easy to locate your exercise files, browse to where the files are located when you save the file. Right click on the left pane and select Add Current Folder.

Exercise 1-7:

Copying a Style to a Template

Drawing Name: Styles1.dwg, Imperial (Metric).ctb.dwt
Estimated Time: 20 minutes

This exercise reinforces the following skills:

- Style Manager
- Use of Templates
- Properties dialog

1.    Close any open drawings.

   Go to the Application menu and select
   Close → All Drawings.

2.   Start a New Drawing using QNEW.

3.   Activate the **Manage** ribbon.

   Style
   Manager | Select the **Style Manager** tool.

4.   Select the **File Open** tool in the Style Manager dialog.

5.   Locate the *styles.dwg* you created.
   Select **Open**.

   Note: Make sure the Files of type is set to dwg or you will not see your files.

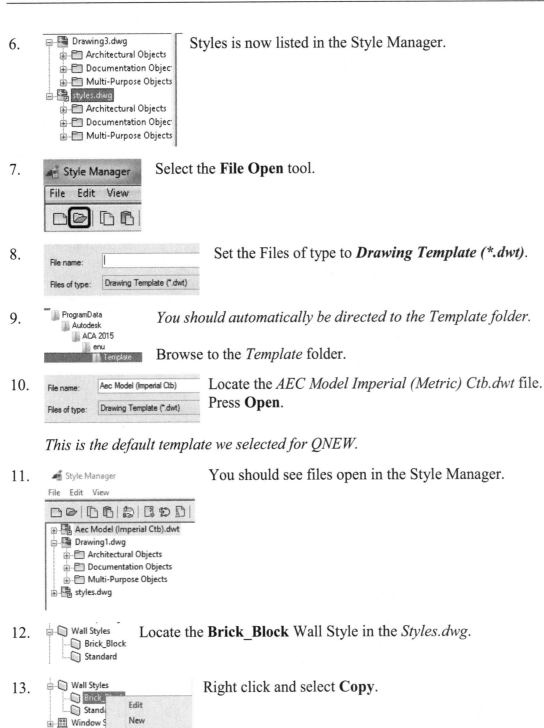

6.   Styles is now listed in the Style Manager.

7.   Select the **File Open** tool.

8.   Set the Files of type to ***Drawing Template (*.dwt)***.

9.   *You should automatically be directed to the Template folder.*

       Browse to the *Template* folder.

10.   Locate the *AEC Model Imperial (Metric) Ctb.dwt* file. Press **Open**.

This is the default template we selected for QNEW.

11.   You should see files open in the Style Manager.

12.   Locate the **Brick_Block** Wall Style in the *Styles.dwg*.

13.   Right click and select **Copy**.

14. 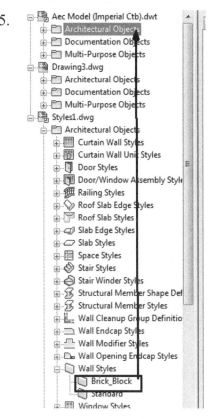 Locate Wall Styles under the template drawing.

Highlight, right click and select **Paste**.

| New |
| Synchronize with F |
| Update Standards 1 |
| Version Styles... |
| Copy |
| Paste |

15. You could also just drag and drop the Brick_block style from the Styles file to the template.

- Aec Model (Imperial Ctb).dwt
  - Architectural Objects
  - Documentation Objects
  - Multi-Purpose Objects
- Drawing3.dwg
  - Architectural Objects
  - Documentation Objects
  - Multi-Purpose Objects
- Styles1.dwg
  - Architectural Objects
    - Curtain Wall Styles
    - Curtain Wall Unit Styles
    - Door Styles
    - Door/Window Assembly Style
    - Railing Styles
    - Roof Slab Edge Styles
    - Roof Slab Styles
    - Slab Edge Styles
    - Slab Styles
    - Space Styles
    - Stair Styles
    - Stair Winder Styles
    - Structural Member Shape Def
    - Structural Member Styles
    - Wall Cleanup Group Definitio
    - Wall Endcap Styles
    - Wall Modifier Styles
    - Wall Opening Endcap Styles
    - Wall Styles
      - Brick_Block
      - Standard
    - Window Styles

16. **Apply** Press **Apply**.

17. The following drawing has changed.
    C:\ProgramData\Autodesk\ACA
    Nautilus\enu\Template\Aec Model (Imperial
    Ctb).dwt

    Do you wish to save changes to this file?

    | Yes | No |

    Press **Yes**.

18. Close the Style Manager.

19. Drawing1*   Look at the drawing folder tabs. The Styles.dwg and the template drawing are not shown. These are only open in the Style Manager.

20. Select the **QNEW** tool.

21.  Activate the Home ribbon.
Select the **Wall** tool.

22. The Properties dialog should pop up when the WallAdd tool is selected.

In the Style drop-down list, note that Brick_Block is available.

If you don't see the wall style in the drop-down list, verify that you set the default template correctly in Options.

23. Press Escape to end the WallAdd command.

24. Close all drawing files without saving.

![Tips Tricks]

A fast way to close all open files is to type CLOSEALL on the command line. You will be prompted to save any unsaved files.

Layer Manager

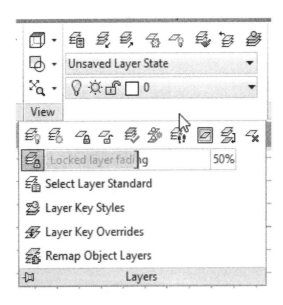

Back in the days of vellum and pencil, drafters would use separate sheets to organize their drawings, so one sheet might have the floor plan, one sheet the site plan, etc. The layers of paper would be placed on top of each other and the transparent quality of the vellum would allow the drafter to see the details on the lower sheets. Different colored pencils would be used to make it easier for the drafter to locate and identify elements of a drawing, such as dimensions, electrical outlets, water lines, etc.

When drafting moved to Computer Aided Design, the concept of sheets was transferred to the use of Layers. Drafters could assign a Layer Name, color, linetype, etc. and then place different elements on the appropriate layer.

AutoCAD Architecture has a Layer Management system to allow the user to implement AIA standards easily.

Layer Manager

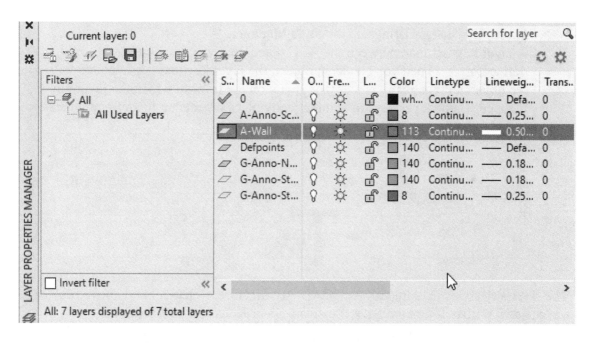

The Layer Manager helps you organize, sort, and group layers, as well as save and coordinate layering schemes. You can also use layering standards with the Layer Manager to better organize the layers in your drawings.

When you open the Layer Manager, all the layers in the current drawing are displayed in the right panel. You can work with individual layers to:

- Change layer properties by selecting the property icons
- Make a layer the current layer
- Create, rename, and delete layers

If you are working with drawings that contain large numbers of layers, you can improve the speed at which the Layer Manager loads layers when you open it by selecting the Layer Manager/Optimize for Speed option in your AEC Editor options.

The Layer Manager has a tool bar as shown.

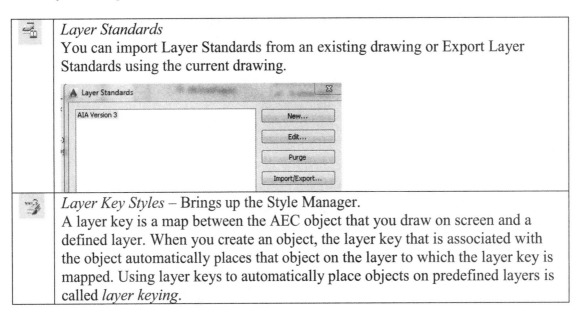	*Layer Standards* You can import Layer Standards from an existing drawing or Export Layer Standards using the current drawing.
	Layer Key Styles – Brings up the Style Manager. A layer key is a map between the AEC object that you draw on screen and a defined layer. When you create an object, the layer key that is associated with the object automatically places that object on the layer to which the layer key is mapped. Using layer keys to automatically place objects on predefined layers is called *layer keying*.

The layer key styles in Autodesk AutoCAD Architecture, Release 3 and above replace the use of LY files. If you have LY files from S8 or Release 1 of AutoCAD Architecture that you want to use, then you can create a new layer key style from an LY file. On the command line enter **-AecLYImport** to import legacy LY files.

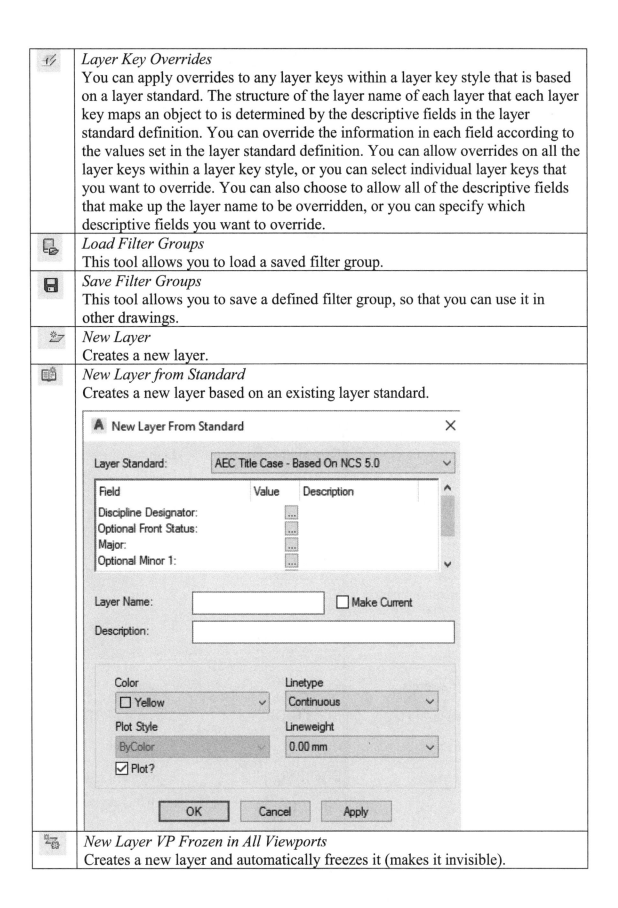	*Layer Key Overrides* You can apply overrides to any layer keys within a layer key style that is based on a layer standard. The structure of the layer name of each layer that each layer key maps an object to is determined by the descriptive fields in the layer standard definition. You can override the information in each field according to the values set in the layer standard definition. You can allow overrides on all the layer keys within a layer key style, or you can select individual layer keys that you want to override. You can also choose to allow all of the descriptive fields that make up the layer name to be overridden, or you can specify which descriptive fields you want to override.
	Load Filter Groups This tool allows you to load a saved filter group.
	Save Filter Groups This tool allows you to save a defined filter group, so that you can use it in other drawings.
	New Layer Creates a new layer.
	New Layer from Standard Creates a new layer based on an existing layer standard.
	New Layer VP Frozen in All Viewports Creates a new layer and automatically freezes it (makes it invisible).

✕	*Delete Layer* Highlight the layer(s) you wish to delete and select this tool.
◌⊣	*Make selected layer current* Highlight a layer name in the list and press this button to set current.

The following are the default layer keys used by Autodesk AutoCAD Architecture when you create AEC objects.

Default layer keys for creating AEC objects

Layer Key	Description	Layer Key	Description
ANNDTOBJ	Detail marks	COMMUN	Communication
ANNELKEY	Elevation Marks	CONTROL	Control systems
ANNELOBJ	Elevation objects	CWLAYOUT	Curtain walls
ANNMASK	Masking objects	CWUNIT	Curtain wall units
ANNMATCH	Match Lines	DIMLINE	Dimensions
ANNOBJ	Notes, leaders, etc.	DIMMAN	Dimensions (AutoCAD points)
ANNREV	Revisions	DOOR	Doors
ANNSXKEY	Section marks	DOORNO	Door tags
ANNSXOBJ	Section marks	DRAINAGE	Drainage
ANNSYMOBJ	Annotation marks	ELEC	Electric
APPL	Appliances	ELECNO	Electrical tags
AREA	Areas	ELEV	Elevations
AREAGRP	Area groups	ELEVAT	Elevators
AREAGRPNO	Area group tags	ELEVHIDE	Elevations (2D)
AREANO	Area tags	EQUIP	Equipment
CAMERA	Cameras	EQUIPNO	Equipment tags
CASE	Casework	FINCEIL	Ceiling tags
CASENO	Casework tags	FINE	Details- Fine lines
CEILGRID	Ceiling grids	FINFLOOR	Finish tags

Layer Key	Description	Layer Key	Description
CEILOBJ	Ceiling objects	FIRE	Fire system equip.
CHASE	Chases	FURN	Furniture
COGO	Control Points	FURNNO	Furniture tags
COLUMN	Columns	GRIDBUB	Plan grid bubbles
Layer Key	**Description**	**Layer Key**	**Description**
GRIDLINE	Column grids	SITE	Site
HATCH	Detail-Hatch lines	SLAB	Slabs
HIDDEN	Hidden Lines	SPACEBDRY	Space boundaries
LAYGRID	Layout grids	SPACENO	Space tags
LIGHTCLG	Ceiling lighting	SPACEOBJ	Space objects
LIGHTW	Wall lighting	STAIR	Stairs
MASSELEM	Massing elements	STAIRH	Stair handrails
MASSGRPS	Massing groups	STRUCTBEAM	Structural beams
MASSSLCE	Massing slices	STRUCTBEAMIDEN	Structural beam tags
MED	Medium Lines	STRUCTBRACE	Structural braces
OPENING	Wall openings	STRUCTBRACEIDEN	Structural brace tags
PEOPLE	People	STRUCTCOLS	Structural columns
PFIXT	Plumbing fixtures	STRUCTCOLSIDEN	Structural column tags
PLANTS	Plants - outdoor	SWITCH	Electrical switches
PLANTSI	Plants - indoor	TITTEXT	Border and title block
POLYGON	AEC Polygons	TOILACC	Arch. specialties
POWER	Electrical power	TOILNO	Toilet tags
PRCL	Property Line	UTIL	Site utilities
PRK-SYM	Parking symbols	VEHICLES	Vehicles
ROOF	Rooflines	WALL	Walls
ROOFSLAB	Roof slabs	WALLFIRE	Fire wall patterning
ROOMNO	Room tags	WALLNO	Wall tags
SCHEDOBJ	Schedule tables	WIND	Windows

SEATNO	Seating tags	WINDASSEM	Window assemblies
SECT	Miscellaneous sections	WINDNO	Window tags
SECTHIDE	Sections (2D)		

Display Manager

Exercise 1-8:

Exploring the Display Manager

Drawing Name: display_manager.dwg
Estimated Time: 15 minutes

This exercise reinforces the following skills:

❑   Display Manager

1.   Open *display_manager.dwg*.

2.   Access the **Display Manager** from the Manage ribbon.

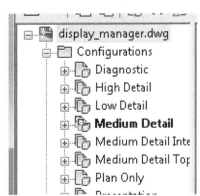

3.   The display system in Autodesk® AutoCAD Architecture controls how AEC objects are displayed in a designated viewport. By specifying the AEC objects you want to display in a viewport and the direction from which you want to view them, you can produce different architectural display settings, such as floor plans, reflected plans, elevations, 3D models, or schematic displays.

4.

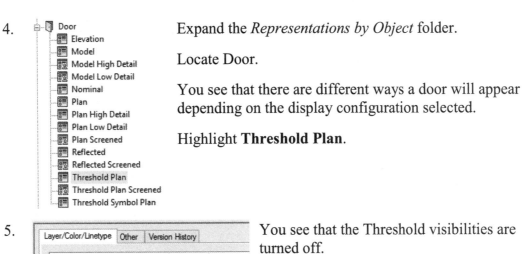

Expand the *Representations by Object* folder.

Locate Door.

You see that there are different ways a door will appear depending on the display configuration selected.

Highlight **Threshold Plan**.

5.

You see that the Threshold visibilities are turned off.

6.

Highlight the **Plan** configuration.

7.

You see that some door components are turned on and others are turned off.

8.

Highlight the Model configuration.

9. You see that some door components are turned on and others are turned off.

10. Go to the Sets folder. The Set highlighted in bold is the active Display Representation.

Highlight **Plan**.

11.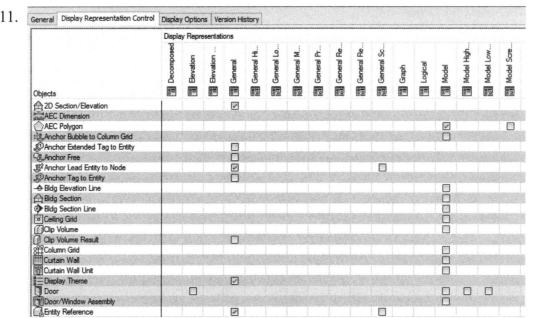

Select the Display Representation Control tab.
Locate **Door** under Objects.
Note that some display representations are checked and others are unchecked.
A check indicates that the door is visible. Which door components are visible is controlled under Representations by Object.

The three different folders – Configurations, Sets, and Representations by Object – are different ways of looking at the same elements. Each folder organizes the elements in a different way. If you change the display setting in one folder, it updates in the other two folders.

12. Expand the **Configurations** folder.

ex1-8.dwg
    Configurations
        Diagnostic
        High Detail
        Low Detail
        Medium Detail
        Medium Detail Intermediate Level
        Medium Detail Top Level
        Plan Only

13. Expand **Medium Detail**.
    You see the different display sets. Note that Medium Detail does not hold all the different display sets. You can add or remove display sets.
    Close the Display Manager.

Medium Detail
    Model
    Section_Elev
    Plan Diagnostic
    Plan

14. Close the drawing without saving.

Exercise 1-9:

Creating a Text Style

Drawing Name: new using QNEW
Estimated Time: 15 minutes

This exercise reinforces the following skills:

- ❑ CUI – Custom User Interface
- ❑ Ribbon
- ❑ Panel
- ❑ Tab

1. Start a new drawing using the QNEW tool.

2.  Select the Text Style tool on the Home ribbon.

   It is located on the Annotation panel under the drop-down.

3.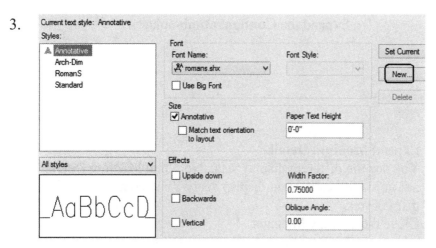

Press the **New** button in the Text Styles dialog box.

4.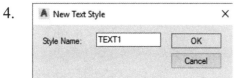

In the New Text Style dialog box, enter **TEXT1** for the name of the new text style and press **OK**.

5.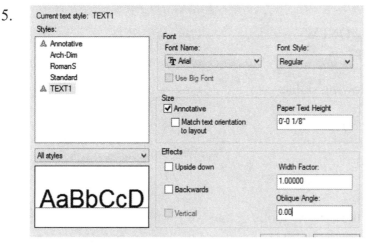

Fill in the remaining information in the Text Style dialog box as shown.

Set the Font Name to **Arial**.
Set the Font Style to **Regular**.
Set the Paper Text Height to **1/8″**.
Set the Width Factor to **1.00**.

6. Press the Apply button first and then the Close button to apply the changes to TEXT1 and then close the dialog box.

7. Save as *fonts.dwg*.

> ➢ You can use PURGE to eliminate any unused text styles from your drawing.
> ➢ You can import text styles from one drawing to another using the Design Center.
> ➢ You can store text styles in your template for easier access.
> ➢ You can use any Windows font to create AutoCAD Text.

Exercise 1-10:

Project Settings

Drawing Name: new using QNEW
Estimated Time: 5 minutes

This exercise reinforces the following skills:

- Project Navigator
- Project Path Options

In order for tags on windows, doors, etc. to work properly, you must assign your building model to a project.

1.   Start a new drawing by pressing the + tab.

2.   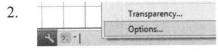   Right click in the command window and select **Options**.

3.      Select the **AEC Project Defaults** tab.

4.   Set the AEC Project Location Search Path to the location where you are saving your files.

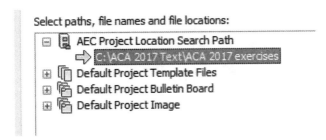

Close the dialog.

5.      Launch the **Project Navigator** from the Quick Access toolbar.

6.   Launch the **Project Browser** using the toolbar on the bottom of the Project Navigator palette.

7. 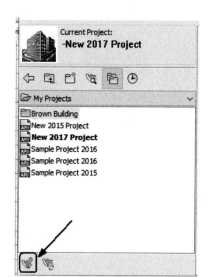 Select the **New Project** tool.

8. Type **A102** for the Project Number.

Type **Brown Residence** for the Project Name.

Enter a project description.

Press **OK**.

9. Close the Project Browser.

10. Save as *ex1-10.dwg*.

Lesson 2:
Site Plans

Most architectural projects start with a site plan. The site plan indicates the property lines, the house location, a North symbol, any streets surrounding the property, topographical features, location of sewer, gas, and/or electrical lines (assuming they are below ground – above ground connections are not shown), and topographical features.

When laying out the floor plan for the house, many architects take into consideration the path of sun (to optimize natural light), street access (to locate the driveway), and any noise factors. AutoCAD Architecture includes the ability to perform sun studies on your models.

AutoCAD Architecture allows the user to simulate the natural path of the sun based on the longitude and latitude coordinates of the site, so you can test how natural light will affect various house orientations.

A plot plan must include the following features:

- Length and bearing of each property line
- Location, outline, and size of buildings on the site
- Contour of the land
- Elevation of property corners and contour lines
- North symbol
- Trees, shrubs, streams, and other topological items
- Streets, sidewalks, driveways, and patios
- Location of utilities
- Easements and drainages (if any)
- Well, septic, sewage line, and underground cables
- Fences and retaining walls
- Lot number and/or address of the site
- Scale of the drawing

The plot plan is drawn using information provided by the county/city and/or a licensed surveyor.

When drafting with pencil and vellum, the drafter will draw to a scale, such as ⅛″ = 1′, but with AutoCAD Architecture, you draw full-size and then set up your layout to the proper scale. This ensures that all items you draw will fit together properly. This also is important if you collaborate with outside firms. For example, if you want to send your files to a firm that does HVAC, you want their equipment to fit properly in your design. If everyone drafts 1:1 or full-size, then you can ensure there are no interferences or conflicts.

Exercise 2-1:

Creating Custom Line Types

Drawing Name: New
Estimated Time: 15 minutes

This exercise reinforces the following skills:

- ❑ Creation of linetypes
- ❑ Customization

Architectural drafting requires a variety of custom linetypes in order to display specific features. The standard linetypes provided with AutoCAD Architecture are insufficient from an architectural point of view. You may find custom linetypes on Autodesk's website, www.cadalog.com, or any number of websites on the Internet. However, the ability to create linetypes as needed is an excellent skill for an architectural drafter.

It's a good idea to store any custom linetypes in a separate file. The standard file for storing linetypes is acad.lin. If you store any custom linetypes in acad.lin file, you will lose them the next time you upgrade your software.

A video of this lesson is available on my MossDesigns channel on Youtube at https://www.youtube.com/watch?v=gSCyFJm6pfY

1. *Once the Express Tools are installed, you can start this exercise. Refer to Lesson 1 for instructions on how to install the Express Tools. If you are unable to install the Express Tools, you can skip this lesson and use the custom file included with the file downloads.*

   Drawing1.dwg

   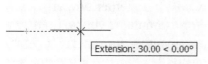

   To verify that the Express Tools are installed, simply look at your ribbon headings.

2. Start a new drawing using **QNEW**.

3. ——————————— ——————— ——————— ————————————

   Draw a property line.
   This property line was created as follows:
       Set ORTHO ON.
       Draw a horizontal line 100 units long.

   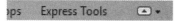

   Extension: 30.00 < 0.00°

   Use SNAP FROM to start the next line @30,0 distance from the previous line. You can also simply use object tracking to locate the next line.
   The short line is 60 units long.

Use SNAP FROM to start the next line @30,0 distance from the previous line.
The second short line is 60 units long.
Use SNAP FROM to start the next line @30,0 distance from the previous line.
Draw a second horizontal line 100 units long.

4.

Activate the Express Tools ribbon.

Select the **Tools** drop-down.

Go to **Express→Tools→Make Linetype**.

5. Browse to the folder where you are storing your work.

Create a file name called *custom-linetypes.lin* and press **Save**.

Place all your custom linetypes in a single drawing file and then use the AutoCAD Design Center to help you locate and load the desired linetype.

---

6. When prompted for the linetype name, type property-line.
When prompted for the linetype description, type property-line.
Specify starting point for line definition; select the far left point of the line type.
Specify ending point for line definition; select the far right point of the line type.
When prompted to select the objects, start at the long line on the left and pick the line segments in order.

7. Type **linetype** to launch the Linetype Manager.

8.

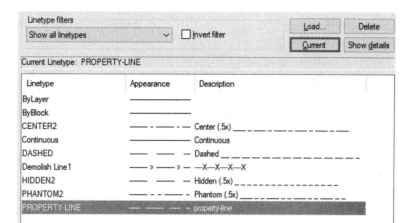

The linetype you created is listed.

Highlight the PROPERTY-LINE linetype and select CURRENT.

Press **OK**.

9. Draw some lines to see if they look OK. Make sure you make them long enough to see the dashes. If you don't see the dashes, adjust the linetype scale using Properties. Set the scale to 0.0125.

10. 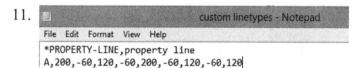 Locate the *custom-linetypes.lin* file you created.

    Open it using NotePad.

11.

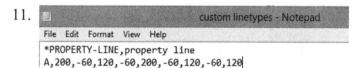

You see a description of the property-line.

12. Save as *ex2-1.dwg*.

Exercise 2-2:

Creating New Layers

Drawing Name: New
Estimated Time: 20 minutes

This exercise reinforces the following skills:

❑ Design Center
❑ Layer Manager
❑ Creating New Layers
❑ Loading Linetypes

1. Start a new drawing using QNEW.

   Open Ex2-1.dwg.

2.  You should see two file tabs for the two drawings that are currently open.
If you don't see the file tabs, they might not be enabled.

Activate the View ribbon and verify that the File Tabs is shaded.

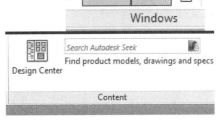

3. Launch the Design Center.
The Design Center is on the Insert ribbon in the Content section. It is also located in the drop-down under the Content Browser. You can also launch using Ctl+2.

4. 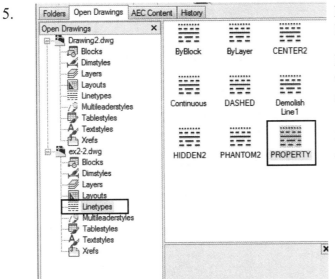 Select the **Open Drawings** tab.

Browse to the *Ex2-1.dwg* file.

Open the Linetypes folder.

Scroll to the **Property-line** linetype definition.

5. Highlight the PROPERTY-LINE linetype.

Drag and drop into your active window or simply drag it onto the open drawing listed.

This copies the property-line linetype into the drawing.

6. Close the Design Center.

7. Activate the **Layer Properties Manager** on the Home ribbon.

8. Note that several layers have already been created. The template automatically sets your Layer Standards to AIA.

9. Select the **New Layer from Standard** tool.

10. 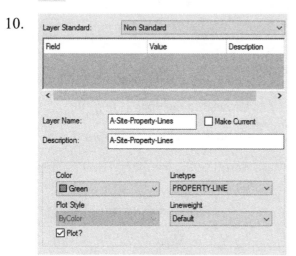 Under Layer Standard, select **Non Standard**.

    Name the layer **A-Site-Property-Lines**.

    Set the Color to **Green**.

    Select the **Property-Line** linetype.

    Set the Lineweight to **Default**.

    Enable **Plot**.
    Press **Apply**.

11. Create another layer.

    Name the layer **A-Contour-Line**.

    Set the color to **42**.

    Select the linetype pane for the new linetype.

    Press **Other**.

12. Enable **Show Linetypes in File**.

    Highlight **DASHEDX2** and press **OK** to assign it to the selected layer name.

13.

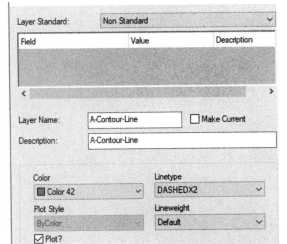

Note that the linetype was assigned properly.

Press **OK**.

14.

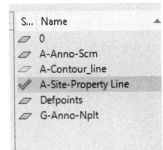

Set the A-Site-Property-Line layer current.

Highlight the layer name then select the green check mark to set current.

You can also double left click to set the layer current.

15. Right click on the header bar and select **Maximize all columns** to see all the properties for each layer.

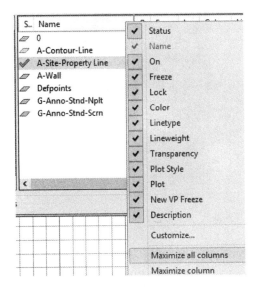

16. Use the X in the upper corner to close the Layer Manager.

17. Save as *ex2-2.dwg*.

Exercise 2-3:
Creating a Site Plan

Drawing Name: Ex2-2.dwg
Estimated Time: 30 minutes

This exercise reinforces the following skills:

- ❑ Use of ribbons
- ❑ Documentation Tools
- ❑ Elevation Marks
- ❑ Surveyors' Angles

1.   Open or continue working in *ex2-2.dwg*.

2.   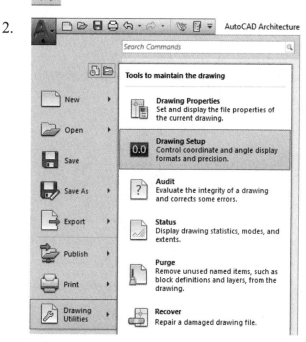   Go to **Drawing Utilities→Drawing Setup**.

This is located on the Application Menu panel.

3.

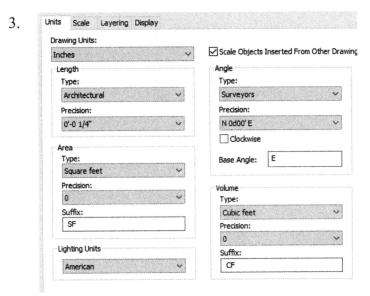

Select the **Units** tab.
Set the Drawing Units to **Inches**.
Set up the Length Type to **Architectural**.
Set the Precision to ¼″.
Set the Angle Type to **Surveyors**.
Set the Precision to **N 0d00′ E**.
Press **Apply**.

4.

You have changed the units for this drawing, which will be used on all new items. What do you want to do?

→ Rescale modelspace and paperspace objects
   Make all match the new units

→ Rescale only modelspace objects
   Make them match the new units

→ Don't rescale any existing objects

You will see this dialog box.

We haven't drawn anything yet, but it is good to know that ACA will automatically rescale any existing elements in a drawing if you change the units.

Select **Don't rescale any existing objects**.

Press **OK** to close the dialog box.

You can set your units as desired and then enable **Save As Default** .
Then, all future drawings will use your unit settings.

5. Draw the property line shown on the **A-Site-Property-Lines** layer.

   Specify first point: 1′,5′
   Specify next point: @186′ 11″ < s80de
   Specify next point: @75′<n30de
   Specify next point: @125′11″<n
   Specify next point: @250′<180

   This command sequence uses
   Surveyor's Units.

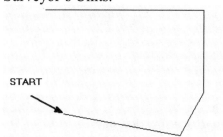

   START

   You can use the TAB key to advance to
   the surveyor unit box.

Metric data entry using centimeters:

```
Specify first point: 12,60
Specify next point or [Undo]: @2243<s80de
Specify next point or [Undo]: @900<n30de
Specify next point or [Close/Undo]: @1511<n
Specify next point or [Close/Undo]: @3000<180
Specify next point or [Close/Undo]:
```

If you double click the mouse wheel, the
display will zoom to fit so you can see
your figure.

Turn ORTHO off before creating the arc.

6. Draw an arc using Start, End, Direction to close
   the figure.

   Arc, 3 Point

   Arc, Start, Center, End

   Arc, Start, Center, Angle

   Arc, Start, Center, Length

   Arc, Start, End, Angle

   Arc, Start, End, Direction

7. 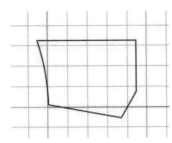 Select the bottom point as the Start point.
Right click and select the End option.
Select the top point at the End point.
Click inside the figure to set the direction of the arc.

8. Set the **A-Contour-Line layer** current.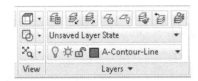

9. 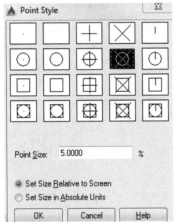 Use the **DIVIDE** command to place points as shown.
The top line is divided into five equal segments.
The bottom-angled line is divided into four equal segments.
The small angled line is divided into two equal segments.

10. 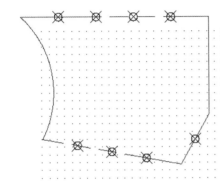 Type **DDPTYPE** to access the Point Style dialog or use the Point Style tool that was added to the Format tab on the ribbon in Lesson 1.

Go to **Format→Point Style** to set the points so they are visible.

Select the Point Style indicated.

Press **OK**.

11. Draw contour lines using **Pline** and **NODE** Osnaps as shown.

Be sure to put the contour lines on the contour line layer.

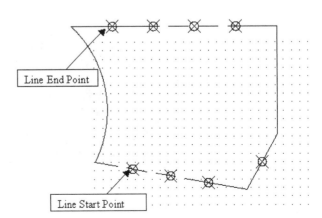

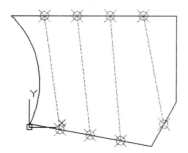

12. Type **qselect** on the *Command* line.

13.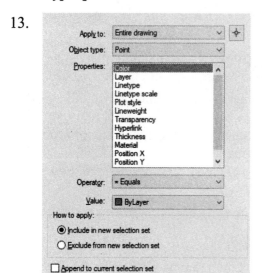

Set the Object Type to **Point**.
Set the Color = Equals **ByLayer**.

Press **OK**.

14.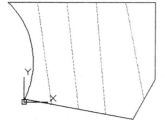

All the points are selected.
Right click and select **Basic Modify Tools→ Delete**.

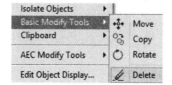

15. Type UCSICON, OFF to turn off the UCS icon.

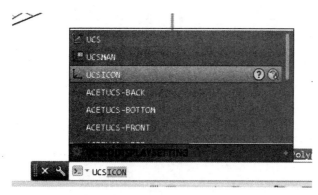

When you start typing in the command line, you will see a list of commands.

Use the up and down arrow keys on your keyboard to select the desired command.

16.  Activate the **Design Tools** palette located under the Tools drop-down on the Home ribbon.

17. 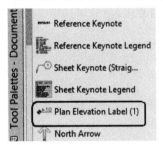 Right click on the gray bar on the palette to activate the short-cut menu.

Left click on **All Palettes** to enable the other palettes.

18. Activate the Annotation tab on the palette.

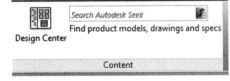 Locate the **Plan Elevation Label**, but it's not the type we need for a site plan.

Instead we'll have to add an elevation label from the Design Center to the Tool Palette.

19. 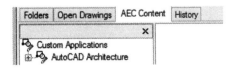 Type **Ctl+2** to launch the Design Center.
Or:
Activate the **Insert** ribbon.
Select the **Design Center** tool under Content.
Or:
Type **DC**.

20. Select the **AEC Content** tab.

21. Browse to
    Imperial/Documentation/Elevation Labels/ 2D Section.

    [Metric/Documentation/Elevation Labels/2D Section].

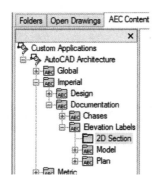

22.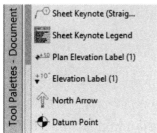

    Locate the Elevation Label (1) file.

    Highlight the elevation label.

    Drag and drop onto the Tool Palette.

    Close the Design Center.

23.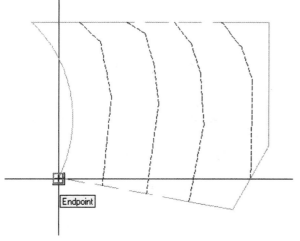

    Select the Elevation Label on the Tool Palette.

    Select the Endpoint shown.

24. Set the Elevation to **0″ [0.00]**.
    Set the Prefix to **EL**.
    Press **OK**.

25.  Zoom in to see the elevation label.

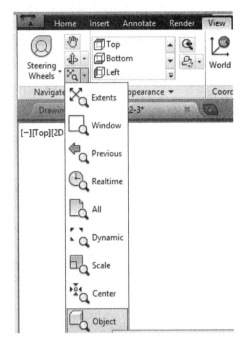 You can use the **Zoom Object** tool to zoom in quickly.

This is located on the View ribbon.

 The Elevation Label is automatically placed on the G-Anno-Dims layer.

To verify this, just select the label and look at the Layers panel on the Home ribbon.

26.  Place an elevation label on the upper left vertex as shown.

27. 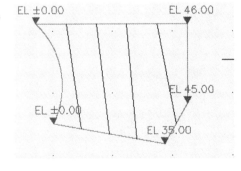 Place labels as shown.

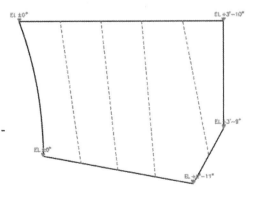

28.

Bound spa...	By style (...
Scale	**−**
X	36.00
Y	36.00
Z	36.00

To change the size of the labels, select and set the Scale to **36.00** on the Properties dialog.

29.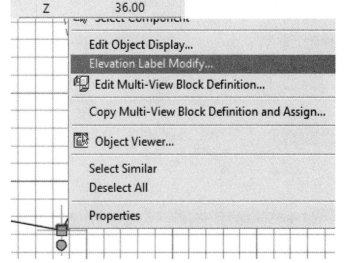

To modify the value of the elevation label, select the label.

Right click and select **Elevation Label Modify**.

30.

Geometry	
Start X	1'-0"
Start Y	5'-0"
Start Z	0"
End X	185'-0 15/16"
End Y	-27'-5 1/2"
End Z	2'-11"
Delta X	184'-0 15/16"
Delta Y	-32'-5 1/2"

To change the toposurface to a 3D surface, select each line and modify the Z values to match the elevations.

31.

Geometry	
Start X	185'-0 15/16"
Start Y	-27'-5 1/2"
Start Z	2'-11"
End X	222'-6 15/16"
End Y	37'-5 15/16"
End Z	3'-9"
Delta X	37'-6"
Delta Y	64'-11 7/16"
Delta Z	10"

32.

Geometry	
Start X	222'-6 15/16"
Start Y	37'-5 15/16"
Start Z	3'-9"
End X	222'-6 15/16"
End Y	163'-4 15/16"
End Z	3'-10"
Delta X	0"
Delta Y	125'-11"
Delta Z	1"
Length	125'-11"
Angle	90.00

33.

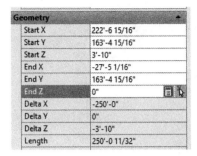

Geometry	
Start X	222'-6 15/16"
Start Y	163'-4 15/16"
Start Z	3'-10"
End X	-27'-5 1/16"
End Y	163'-4 15/16"
End Z	0"
Delta X	-250'-0"
Delta Y	0"
Delta Z	-3'-10"
Length	250'-0 11/32"

34.    Use the ViewCube to inspect your toposurface and then return to a top view.

35. Open the *fonts.dwg*. This is the file you created in a previous exercise.

36. Type **Ctl+2** to launch the Design Center.

37. Browse to the Textstyles folder under the fonts.dwg.

38.    Drag and drop the **Text1** text style into the drawing.

Close the Design Center.

39.   Set **G-Anno-Dims** as the current Layer.

40. 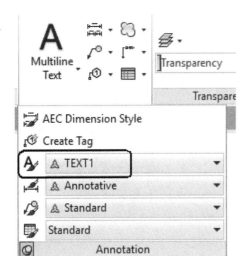   Set the Current Text Style to **TEXT1.**

41. 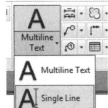   Select the **Single Line** text tool from the Home ribbon.

42.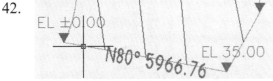

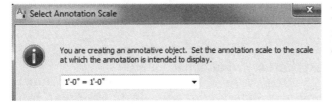

If this dialog appears, set the scale to the same scale as the current view.

Select the midpoint of the line as the insertion point.
Right click and select **Justify**.
Set the Justification to **Center**.
Set the rotation angle to **–10** degrees.
Use %%d to create the degree symbol.

43.    To create the S60… note, use the **TEXT** command.

Use a rotation angle of 60 degrees.

When you are creating the text, it will preview horizontally. However, once you left click or hit ESCAPE to place it, it will automatically rotate to the correct position.

44.  Create the **Due South 125′ 11″ [Due South 3840.48]** with a rotation angle of 90 degrees.

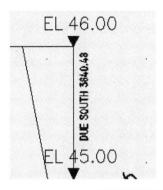

Add the text shown on the top horizontal property line.

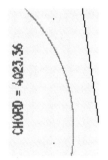

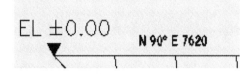

45. Add the Chord note shown. Rotation angle is 90 degrees.

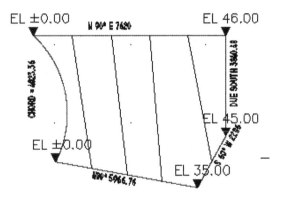

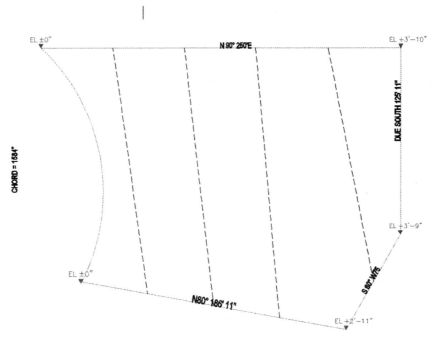

46. 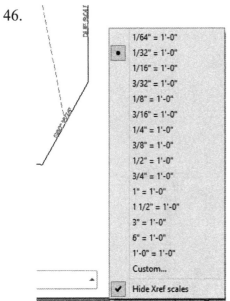   Adjust the scale to **1/32″ = 1′-0″** to see the property line labels.

47. Save the file as *ex2-4.dwg*.

Exercise 2-4:
Creating a Layer User Group

Drawing Name: Ex 2-4.dwg
Estimated Time: 15 minutes

This exercise reinforces the following skills:

- ❑ Use of toolbars
- ❑ Layer Manager
- ❑ Layer User Group

A Layer User Group allows you to group your layers, so you can quickly freeze/thaw them, turn them ON/OFF, etc. *We can create a group of layers that are just used for the site plan and then freeze them when we don't need to see them.*

The difference between FREEZING a Layer and turning it OFF is that entities on FROZEN Layers are not included in REGENS. Speed up your processing time by FREEZING layers. You can draw on a layer that is turned OFF. You can not draw on a layer that is FROZEN.

1.   Open *ex2-4.dwg*.

2.   Select the **Layer Properties Manager** tool from the Home ribbon.

3.   Highlight the word **All** in the Filters pane.

    Right click and select the **New Group Filter**.

4.   Name the group **Site Plan**.

5.

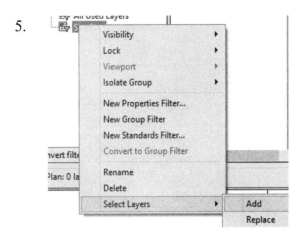

Highlight the Site Plan group.
Right click and select **Select Layers →
Add**.

6. Type **ALL** on the command line.
   This selects all the items in the drawing.
   Press **ENTER** to finish the selection.
   The layers are now listed in the Layer Manager under the Site Plan group.

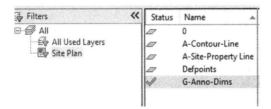

Close the Layer Manager.

7. We can use this group to quickly freeze, turn off, etc. the layers in the Site Plan group.

Save the drawing as *ex2-5.dwg* and close.

QUIZ 1

True or False

1. Doors, windows, and walls are inserted into the drawing as objects.

2. The Architectural Toolbars are loaded from the ACA Menu Group.

3. To set the template used by the QNEW tool, use Options.

4. If you start a New Drawing using the 'QNew' tool shown above, the Start-Up dialog will appear unless you assign a template.

5. Mass elements are used to create Mass Models.

6. The Layer Manager is used to organize, sort, and group layers.

7. Layer Standards are set using the Layer Manager.

8. You must load a custom linetype before you can assign it to a layer.

Multiple Choice
Select the best answer.

9. If you start a drawing using one of the standard templates, AutoCAD Architecture will automatically create _____ layouts.

    A. A work layout plus Model Space
    B. Four layouts, plus Model Space
    C. Ten layouts, plus Model Space
    D. Eleven layouts, plus Model Space

10. Options are set through:

    A. Options
    B. Desktop
    C. User Settings
    D. Template

11. The Display Manager controls:

    A. How AEC objects are displayed in the graphics window
    B. The number of viewports
    C. The number of layout tabs
    D. Layers

12. Mass Elements are created using the _____ Tool Palette.

    A. Massing
    B. Concept
    C. General Drafting
    D. Documentation

13. Wall Styles are created by:

    A. Highlighting a Wall on the Wall tab of the Tool Palette, right click and select Wall Styles
    B. Type **WallStyle** on the command line
    C. Go to **Format→Style Manager**
    D. All of the above

14. To create a New Layer, use:

    A. Type **LayerManager** at the command line
    B. Type **Layer** at the command line
    C. Use **Format→Layer Management→Layer Manager**
    D. All of the above

15. You can change the way AEC objects are displayed by using:

    A. Edit Object Display
    B. Edit Display Properties
    C. Edit Entity Properties
    D. Edit AEC Properties

ANSWERS:

1) T; 2) T; 3) T; 4) F; 5) T; 6) T; 7) T; 8) T; 9) A; 10) A; 11) A; 12) A; 13) D; 14) D; 15) A

Lesson 3:
Floor Plans

AutoCAD Architecture comes with 3D content that you use to create your building model and to annotate your views. In ACA 2018, you may have difficulty locating and loading the various content, so this exercise is to help you set up ACA so you can move forward with your design.

The Content Browser lets you store, share, and exchange AutoCAD Architecture content, tools, and tool palettes. The Content Browser runs independently of the software, allowing you to exchange tools and tool palettes with other Autodesk applications.

The Content Browser is a library of tool catalogs containing tools, tool palettes, and tool packages. You can publish catalogs so that multiple users have access to standard tools for projects.

ACA comes with several tool catalogs. When you install ACA, you enable which catalogs you want installed with the software. By default, Imperial, Metric, and Global are enabled. The content is located in the path: C:\ProgramData\Autodesk\ACA 2018\enu\Tool Catalogs.

Exercise 3-1:
Adding ACA Catalogs to the Catalog Browser

Drawing Name: New
Estimated Time: 10 minutes

This exercise reinforces the following skills:

- ❑ ACA Catalog Browser
- ❑ Adding Catalogs to the Content Browser

1.
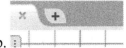
Start a new drawing using QNEW or select the + tab.

2.
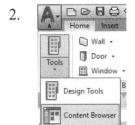
Launch the **Content Browser** – located on the Home ribbon under the Tools drop-down.

3.  Select the Add or Create Catalog tool located on the lower left corner of the Browser dialog.

4. 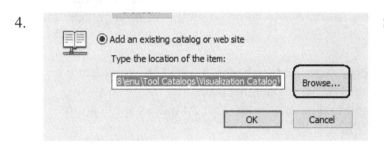 Select the **Browse** button.

5. Browse to the Design – Imperial folder located under:

C:\ProgramData\Autodesk\ACA 2018\enu\Tool Catalogs

6. Select the *Design – Imperial.atc* file.

Press **Open**.

7. Press **OK**.

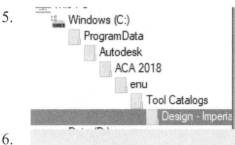

8. The catalog is now listed in the Content Browser.

Design Tool Catalog - Imperial

9. Repeat to add the Design Tool – Catalog – Metric, the Global Catalog, and the Visualization Catalog.

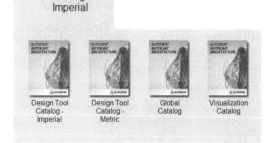

Exercise 3-2:

Adding Tools from the Content Browser to the Tool Palette

Drawing Name: New
Estimated Time: 10 minutes

This exercise reinforces the following skills:

- ❑ ACA Catalog Browser
- ❑ Adding Tools to the Tool Palette

1.  Start a new drawing using QNEW or select the + tab.

2.  Launch the **Content Browser** – located on the Home ribbon under the Tools drop-down.

3.  Launch the **Design Tools** palette– located on the Home ribbon under the Tools drop-down.

4. 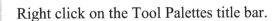 Right click on the Tool Palettes title bar.

    Select **New Palette**.

5. 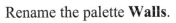 Rename the palette **Walls**.

6.  Left click on the **Design Tool Catalog – Imperial** on the Content Browser to open.

7.  Scroll down the list of categories until you see Walls.

Select **Walls**.

8. In the Search field, type in **Stud-4**.
Press **Go**.

9.  Locate the **Stud-4 Rigid-1.5 Air-1 Brick-4** Wall Style.

Place your cursor over the wall style.
Hold down the left mouse to fill up the eyedropper.
Then place the cursor over the Walls palette and release the left mouse button.

10.  The wall style is added to the Walls tool palette.

11. Locate the **Stud-4 GWB-0.625-2 Layers Each Side** Wall Style.
Place your cursor over the wall style.
Hold down the left mouse to fill up the eyedropper.
Then place the cursor over the Walls palette and release the left mouse button.

12. The wall style is added to the Walls tool palette.

13. Type **CMU-8** in the Search field.

14. Locate the **CMU-8 Rigid 1.5 Air 2 Brick-4** Wall Style.
Place your cursor over the wall style.
Hold down the left mouse to fill up the eyedropper.
Then place the cursor over the Walls palette and release the left mouse button.

15. The Walls tool palette should have three wall styles available.

The floor plan is central to any architectural drawing. In the first exercise, we convert an AutoCAD 2D floor plan to 3D. In the remaining exercises, we work in 3D.

A floor plan is a scaled diagram of a room or building viewed from above. The floor plan may depict an entire building, one floor of a building, or a single room. It may also include measurements, furniture, appliances, or anything else necessary to the purpose of the plan.

Floor plans are useful to help design furniture layout, wiring systems, and much more. They're also a valuable tool for real estate agents and leasing companies in helping sell or rent out a space.

Exercise 3-3:

Going from a 2D to 3D Floor plan

Drawing Name: New
Estimated Time: 45 minutes

This exercise reinforces the following skills:

- ❑ Create Walls
- ❑ Wall Properties
- ❑ Wall Styles
- ❑ Style Manager
- ❑ Insert an AutoCAD drawing
- ❑ Trim, Fillet, Extend Walls

1.  Start a new drawing using QNEW or select the + tab.

2. Type **UNITS**.

   Set the Units to **Inches**.

   Set the Type to **Architectural.**

   Set the Precision to ¼".

   Press **OK**.

3. Since we haven't placed anything in the drawing yet, you can select Option 3 – **Don't rescale any existing objects.**

4.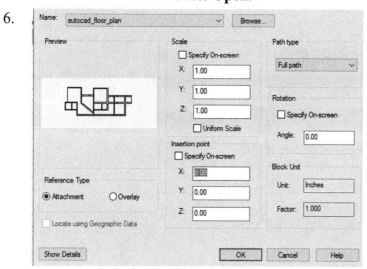

Activate the **Insert** ribbon.

Select **Attach**.

5. | | |
|---|---|
| File name: | autocad_floor_plan |
| Files of type: | Drawing (*.dwg) |

Locate the *autocad_floor_plan.dwg* file in the exercises. *Set your Files of type to Drawing (*.dwg) to locate the file.* Press **Open**.

6.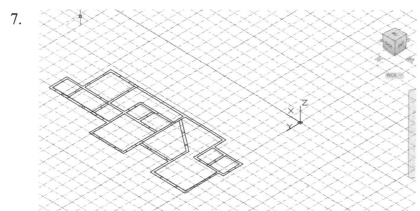

Uncheck Insertion Point.
Uncheck Scale.
Uncheck Rotation.
This sets everything to the default values.

Press **OK**.

7.

Use the ViewCube to switch to a 3D view.

Note that the AutoCAD file is 2D only.

Return to a top view.

8. | |
|---|
| Reload |
| Detach |
| Bind ▶ |
| Clip Xref |
| Frame |

Insert
Bind

Select the attached xref.
Right click and select **Bind→Insert**.
This converts the xref to an inserted block.

9. Select the block reference and type **EXPLODE** to convert to lines.

10.  Activate the **Home** ribbon.

   Select the **Measure** tool on the Inquiry panel.

11. 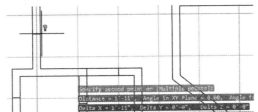 Measure a wall thickness.

   Note that the walls are 1'-11" thick.

12. Launch the **Design Tools** palette from the Home ribbon.

13.  Activate the **Walls** palette.

   This palette was created in the Exercise 3-2.

14.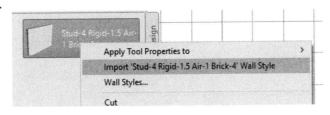

   Locate the Stud-4 Rigid-1.5 Air-1 Brick-4 wall style on the tool palette.
   Right click and select **Import Stud-4 Rigid-1.5 Air-1 Brick-4 Wall Style**.

   This adds the wall style to the active drawing.

   Right click and select **Import Wall Style.**

15. Locate the Stud-4 Rigid-1.5 Air-1 Brick-4 wall style.

   This loads the wall style into the file.

16.

Right click on the **Stud-4 Rigid-1.5 Air-1 Brick-4** wall style and select **Wall Styles**.

This launches the Style Manager.

17.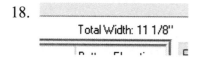

Note that the only wall styles available are Standard and the style that was just imported.

Highlight the **Stud-4 Rigid-1.5 Air-1 Brick-4** wall style.

18. Total Width: 11 1/8"

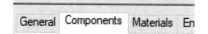

Activate the **Components** tab.

The components tab lists the materials used in the wall construction.

Note the components listed in the Style Manager for the wall style. The total wall thickness is 11-1/8″. We need a wall style that is 1′–11″. We need to add 11- 7/8" of material to the wall style.

19.

Index	Name	Priority	Width	Edge Offset	Function	Dimension	
1	Brick Veneer	810	4"	2 1/2"	Non-Structural	☑	☐
2	Air Gap	700	1"	1 1/2"	--	☐	☐
3	Rigid Insulation	600	1 1/2"	0"	Non-Structural	☐	☐
4	Stud	500	4"	-4"	Structural	☑	☐
5	GWB	1200	5/8"	-4 5/8"	Non-Structural	☐	☐

Highlight the row that lists the GWB material.

GWB stands for Gypsum Wallboard.

20. Select the **Add Component** tool.

21.

Index	Name	Priority	Width	Edge Offset	Function	Dimens
1	Brick Ven...	810	4"	2 1/2"	Non-Struct...	☑
2	Air Gap	700	1"	1 1/2"	--	☐
3	Rigid Insul...	600	1 1/2"	0"	Non-Struct...	☐
4	Stud	500	3 1/2"	-3 1/2"	Structural	☑
5	GWB	1200	5/8"	-4 1/8"	Non-Struct...	☐
6	GWB	1200	5/8"	-4 3/4"	Non-Struct...	☐

Another 5/8″ piece of GWB (gypsum board) is added.

22. Total Width: 11 1/8"   Note that the thickness of the wall updated to: **11 1/8"**.

23.

	Name	Priority	Width	Edge Offset	F
Index					
1	Brick Veneer	810	4"	2 1/2"	N
2	Air Gap	700	1"	1 1/2"	–
3	Rigid Insulation	600	1 1/2"	0"	N
4	Stud	500	4"	-4"	S
5	GWB	1200	5/8"	-4 5/8"	N
6	GWB	1200	5/8"	-4 5/8"	N

Highlight the Brick Veneer material in the top row.

24. Select the **Add Component** tool.

25.

Total Width: 1'-11"

Index	Name	Priority	Width	Edge Offset	Function	Dime
1	Brick Veneer	810	4"	2 1/2"	Non-Structural	☑
2	CMU	810	1' 3.875"	2 1/2"	Non-Structural	☑
3	Air Gap	700	1"	1 1/2"	–	☐
4	Rigid Insulation	600	1 1/2"	0"	Non-Structural	☐
5	Stud	500	4"	-4"	Structural	☑
6	GWB	1200	5/8"	-4 5/8"	Non-Structural	☐
7	GWB	1200	5/8"	-4 5/8"	Non-Structural	☐

Change the name of the second row material to **CMU**.
Set the width to 1' 3.875" thick.

To change the values, just place the cursor in that cell and start typing.

26. Verify that your layers are set as shown.
Verify that the total width is 1' 11".

27. Press **OK** to close the Styles Manager dialog.

28.

Right click on the **Stud-4 Rigid-1.5 Air-1 Brick-4** wall style and select **Apply Tool Properties to → Linework**.

29.

Select the outside segments of the walls.

Do not select any of the interior walls.

Press ENTER when you are done selecting lines.

30. You will be prompted if you want to erase any of the linework. Enter **NO**.

31.

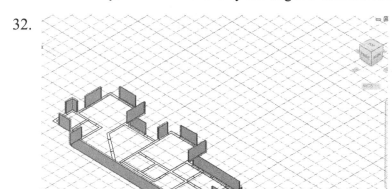

Zoom into one of the walls that was placed.

Note that it is the correct width. The blue arrow indicates the exterior side of the wall. If the blue arrow is inside the building, click on the blue arrow and it will flip the orientation of the wall.

If necessary, move walls so they are aligned with the floor plan's walls.

32.

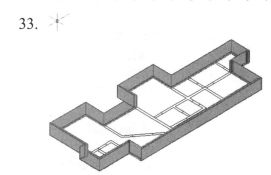

Switch to a 3D view.

You should see 3D walls where you selected lines.

33.

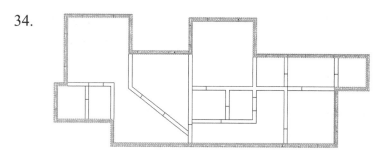

To join the walls together, use FILLET with an R value of 0.
Type FILLET, then select the two walls to be joined to form a corner.

34.

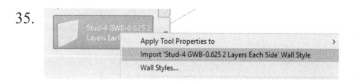

In the plan view, the exterior walls should form a closed figure.

35.

Stud-4 GWB-0.625 2
Layers Each

Apply Tool Properties to ›
Import 'Stud-4 GWB-0.625 2 Layers Each Side' Wall Style
Wall Styles...

Locate the **Stud-4 GWB-0.625-2 Layers Each Side** Wall Style.

Right click and select Import Wall Style.

This loads the wall style into the file.
Right click and select Wall Styles.
This will launch the Styles Manager.

36. Highlight the **Stud-4 GWB-0.625-2 Layers Each Side** Wall Style in the Style Manager list.

Total Width: 6 1/2"

Index	Name	Priority	Width	Edge Offset	Function	Dime ⊞
1	GWB	1210	5/8"	5/8"	Non-Structural	☐
2	GWB	1200	5/8"	0"	Non-Structural	☐
3	Stud	500	4"	-4"	Structural	☑
4	GWB	1200	5/8"	-4 5/8"	Non-Structural	☐
5	GWB	1210	5/8"	-5 1/4"	Non-Structural	☐

Select the **Components** tab.

The total width for this wall style is 6 1/2".

37.

Total Width: 1'-11"

Index	Name	Priority	Width	Edge Offset	Function	Dime ⊞
1	GWB	1210	5/8"	5/8"	Non-Structural	☐
2	GWB	1200	5/8"	0"	Non-Structural	☐
3	Stud	500	1' 9.75"	-4"	Structural	☑
4	GWB	1200	5/8"	-4 5/8"	Non-Structural	☐
5	GWB	1210	5/8"	-5 1/4"	Non-Structural	☐

Change the Stud width to **1' 9 3/4"**.

Adjust the positions of the components so that the wall looks proper.

Press **OK** to close the Style Manager.

38. Select the **Stud-4 GWB-0.625-2 Layers Each Side** wall style.

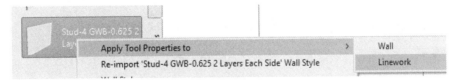

Right click and select **Apply Tool Properties to → Linework**.

39.

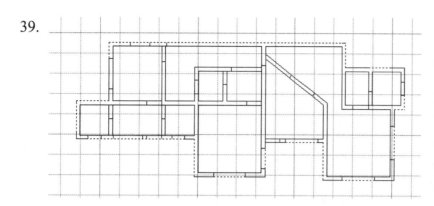

Select the inside segments of the walls.

Do not select any of the exterior walls.

Press ENTER when you are done selecting lines.

40. You will be prompted if you want to erase any of the line work. Enter **NO**.

41.

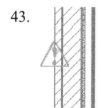

Zoom into one of the walls that was placed. Note that it is the correct width.

The blue arrow indicates the exterior side of the wall. If the blue arrow is inside the building, click on the blue arrow and it will flip the orientation of the wall. Because these are interior walls with gypsum board on both sides, the orientation doesn't matter.

If necessary, move walls so they are aligned with the floor plan's walls.

42.

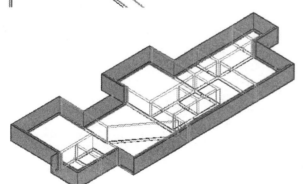

Use the TRIM, EXTEND, and FILLET tools to edit the interior walls.

43.

Some of your walls may display a warning symbol.

This means that you have walls overlapping each other.

Check to see if you have more than one wall or if you need to trim the walls.

44. Save as *ex3-3.dwg*.

The ex3-3 file can be downloaded from the publisher's website, so you can check your file against mine and see how you did.

Exercise 3-4:

Importing a PDF into ACA

Drawing Name: New
Estimated Time: 10 minutes

This exercise reinforces the following skills:

- ❑ Import PDF
- ❑ Create Walls
- ❑ Wall Properties
- ❑ Wall Styles
- ❑ Model and Work space

1. Go to the Application
    Menu (the Capitol Letter
    A).

    Select **New→Drawing**.

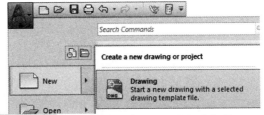

2.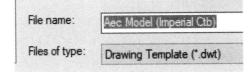

    Select the *Aec Model (Imperial Ctb)* template.

    Press **Open**.

    Note this template uses Architectural units.

3.

    Activate the **Insert** ribbon.

    Select the **Import** tool (located in the middle of the ribbon).

4.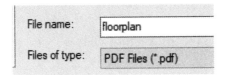

    Select the *floorplan.pdf* file.

5.

You will see a preview of the pdf file.

Press **Open**.

6.

Uncheck specify insertion point on-screen.

This will insert the pdf to the 0,0 coordinate.

Set the Scale to **1200**.

This will scale the pdf.

Set the rotation to 0.

Enable Vector Geometry.
Enable Solid Fills.
Enable TrueType Text.
This will convert any text to AutoCAD text.
Enable Join line and arc segments.
Enable Convert solid fills to hatches.
Enable Apply lineweight properties.
Enable Use PDF layers.

Press OK.

7.

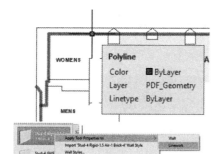

Notice if you hover your mouse over any of the elements imported, they have been converted to ACA elements.

8.

Highlight the **Stud-4 Rigid** wall style on the tool palette.
Right click and select **Apply Tool Properties to Linework.**

9.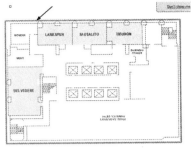

Select the outside polyline on the floorplan.

When prompted to erase existing lines, select **No**.

10.

Highlight the **Stud 4- GWB** wall style.
Right click and select **Apply Tool Properties to Linework.**

When prompted to erase existing lines, select **No**.

11.

Use the FILLET, TRIM, and EXTEND tools to place the interior walls.

12. Save as *ex3-4.dwg*.

You can compare your drawing with mine and see how you did.

Exercise 3-5:
Creating Walls

Drawing Name: New
Estimated Time: 10 minutes

This exercise reinforces the following skills:

- ❏ Create Walls
- ❏ Wall Properties
- ❏ Wall Styles
- ❏ Model and Work space

1.   Go to the Application Menu (the Capitol Letter A).

    Select **New→Drawing**.

2.   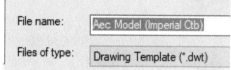

Select the *Aec Model (Imperial Ctb)* template.

Press **Open**.

Note this template uses Architectural units.

3.   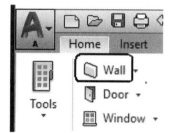   Select the **Wall** tool from the Home ribbon.

4.   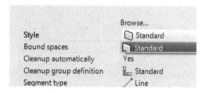

In the Properties dialog, check under the Style drop-down list.

Only the Standard style is available.

This is the wall style that is loaded in the template.

5.   Exit out of the command by pressing ESC.

6.  Launch the Design Tools palette from the Home ribbon.

7. Select the Walls palette.

8. 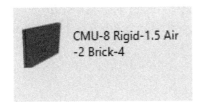 Select the **CMU-8 Rigid-1.5 Air 2 Brick-4** wall style.

9. Toggle **ORTHO** ON.

Start the wall at 0,0.
Create a rectangle 72 inches [1830 mm] tall and 36 inches [914 mm] wide.

You can use Close to close the rectangle.

Place the walls as if you are drawing lines.

10. Go to the **View** ribbon.
11.  Toggle on the Layout tabs.

12. Select the **Work** tab now visible in the lower left corner of the screen.

13. 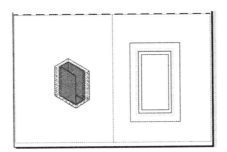 The work tab opens up a layout with two viewports. One viewport is 3D and the other viewport is a top view.

You see that the walls you placed are really 3-dimensional.

14. 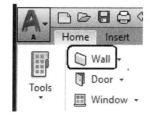 Switch back to the Model space tab.

15. 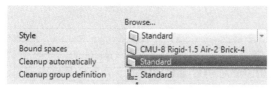 Select the **Wall** tool from the Home ribbon.

16. In the Properties dialog, check under the Style drop-down list.

Note that the CMU wall style is now available under the drop-down list.

17. Exit out of the command by pressing ESC.

18. Save your drawing as e*x3-5.dwg*.

Tips & Tricks

> ➤ If you draw a wall and the materials composing the wall are on the wrong side, you can reverse the direction of the wall. Simply select the wall, right click and select the Reverse option from the menu.

> ➤ To add a wall style to a drawing, you can import it or simply create the wall using the Design Tools.

> ➤ Many architects use external drawing references to organize their projects. That way, teams of architects can concentrate just on their portions of a building. External references also use fewer system resources.

> ➤ You can convert lines, arcs, circles, or polylines to walls. If you have created a floor plan in AutoCAD and want to convert it to 3D, open the floor plan drawing inside of AutoCAD Architecture. Use the Convert to Walls tool to transform your floor plan into walls.

> ➤ To create a freestanding door, press the ENTER key when prompted to pick a wall. You can then use the grips on the door entity to move and place the door wherever you like.

> ➤ To move a door along a wall, use Door→Reposition→Along Wall. Use the OSNAP From option to locate a door a specific distance from an adjoining wall.

Exercise 3-6:

Creating a Floor Plan Using an Image

Drawing Name: new.dwg
Estimated Time: 60 minutes

This exercise reinforces the following skills:

- ❑ Insert Image
- ❑ Add Wall

1. Go to the Application
    Menu (the Capitol Letter
    A).

    Select **New→Drawing**.

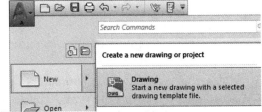

2. 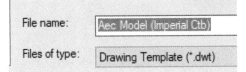   Select the *Aec Model (Imperial Ctb)* template.

    Press **Open**.

    Note this template uses Architectural units.

3. 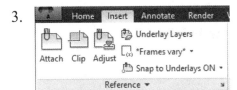   Select the **Insert** ribbon.

    Select the **Attach** tool.

4. File name: |
    Files of type: All image files

    Browse to the folder where the exercises are stored.

    Change the Files of type to **All image files**.

5. File name: floorplan1
    Files of type: All image files

    Select the *floorplan1* file.

    Press **Open**.

6.

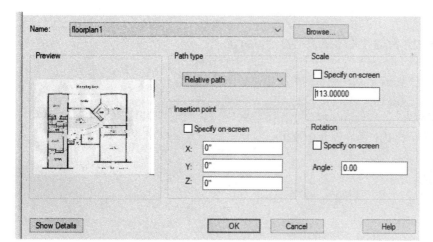

Uncheck the insertion point to insert the image at **0, 0, 0**.

Set the Scale to **113.00**.
Set the Angle to **0.0**.

Press **OK**.

7.

To prevent your image from moving around:

Create a new layer called image.
Select the image.
Right click and select Properties.
Assign the image to the image layer.
Lock the image layer.

8.

Open the Design Tools palette.
Select the Walls palette.

9.

Stud-4 Rigid-.5 Air-1 Bric...

Locate the **Stud-4 Rigid 1.5 Air-1 Brick-4** wall style.

10.

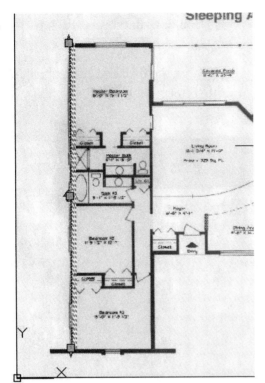

Draw a wall on the far left side of the floor plan, tracing over the wall shown in the image file.

Orient the wall so the exterior side of the wall is on the outside of the building.

11.

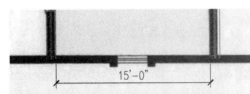

Offset the wall 15' 11-1/8" to the right.

The additional offset takes into account the wall thickness of 11-1/8".

Flip the wall orientation so the wall exterior is on the outside of the building.

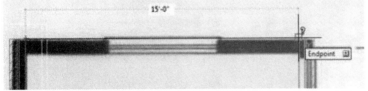

Check the offset distance to ensure the two walls are 15' apart from inside finish face to inside finish face.

12. Trace a horizontal wall using the **Stud-4 Rigid 1.5 Air-1 Brick-4** wall style.

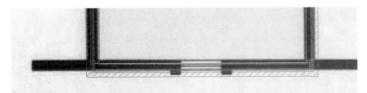

13.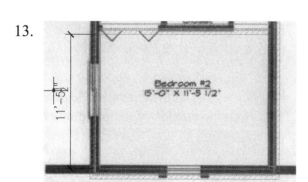

Offset the horizontal wall 12' 4.625".
This is 11' 5 1/2" plus 11 1/8".

Verify that the distance from finish face to finish face is 11' 5 ½".

14. Locate the **Stud-4 GWB-0.625 Each Side** wall style on the Walls palette.

15. Right click and select **Apply Tool Properties to → Wall**.

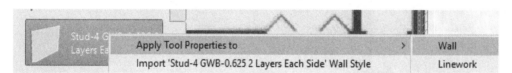

16. Select the upper horizontal wall.

Press **ENTER**.

17.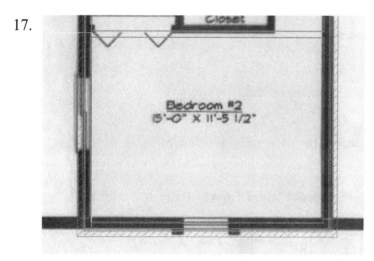

The wall style will update.

18.

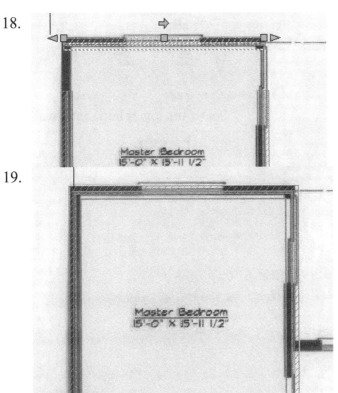

Place a **Stud-4 Rigid 1.5 Air-1 Brick-4** wall at the top horizontal location of the Master Bedroom.

Verify that the orientation is for the exterior side of the wall outside the building.

19.

Use the FILLET command to create corners between the vertical and horizontal walls.

Type FILLET and select the horizontal wall, then select a vertical wall. Repeat for the other side.

20.

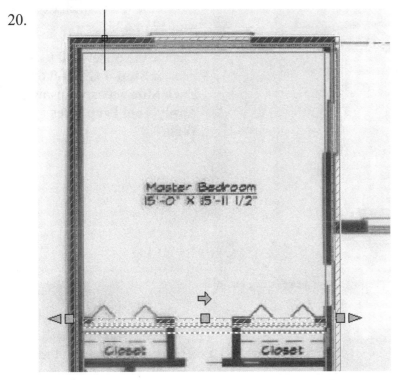

Offset the top horizontal wall 16' 10.625".

21. Stud-4 GWB-0.625 Each Side   Locate the **Stud-4 GWB-0.625 Each Side** wall style on the palette.

22. Right click and select **Apply Tool Properties to → Wall**.

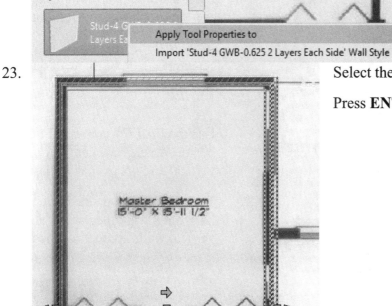

23. 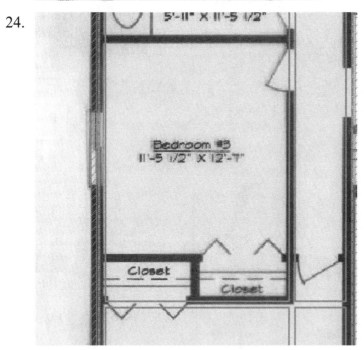 Select the lower horizontal wall.

Press **ENTER**.

24. Offset the left vertical exterior wall **12′ 4.625″**.

Change the offset wall to the interior **Stud-4 GWB-0.625 Each Side** wall style using the **Apply Tool Properties to → Wall**.

25. 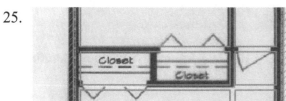 Use an offset of 2′ 0″ to create the closet space.

26.

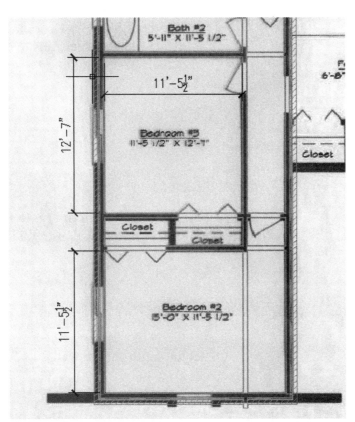

Adjust the position of the walls as needed to ensure they match the floor plan image.

27. Stud-4 Rigid-1.5 Air-1 Brick-4

Select the **Stud-4 Rigid 1.5 Air-1 Brick-4** wall tool from the Design Tools palette.

28.

Segment type	/ Line
Dimensions	
A Width	11 1/8"
B Base height	10'-0"
C Length	1"
Justify	Center
* Offset	0"

On the Properties palette, set the Justify option to **Center**.

29. Trace the remaining south walls of the floor plan.

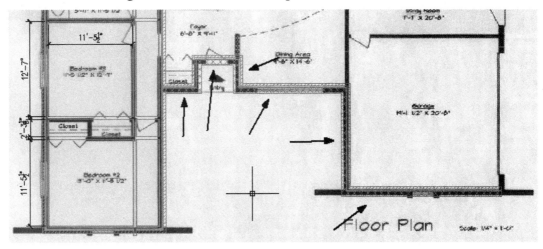

30.

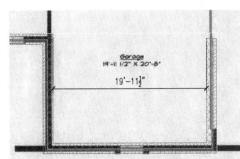

Use the flip arrows to orient the exterior side of the walls to the outside of the building.

31.

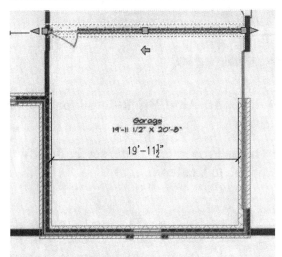

Offset the left garage wall 21′ 3″.

Verify that the dimension from face to face of the interior side of the walls is 19′ 11 ½″.

32.

Offset the south garage wall 20′ 8″.

33.

Select the north garage wall.

In the Properties palette:
Change the wall style to **Stud-2.5 GWB-0.625 Each Side**.

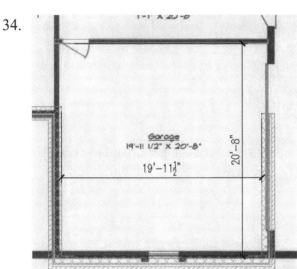

34.

Adjust the position of the garage walls so the distance from interior face to interior face north-south is 20′ 8″ and the distance from interior face to interior face west-east is 19′ 11½″.

35.

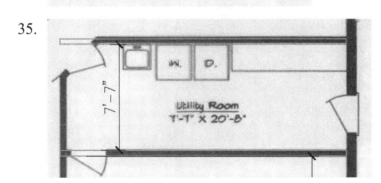

Offset the south utility room wall up 8′ ¼″.

Verify that the distance from interior face to interior face is 7′ 7″.

36.

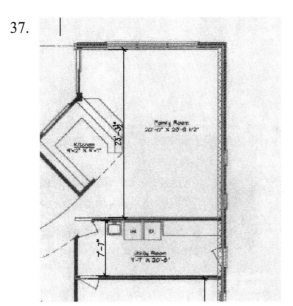

Offset the north utility wall 23′ 9.1325″.

Assign the top wall to the **Stud-4 Rigid 1.5 Air-1 Brick-4** wall style.

Verify that the distance from interior face to interior face is 23′ 3½″.

37.

Use the FILLET command to create the northeast corner of the building.

38.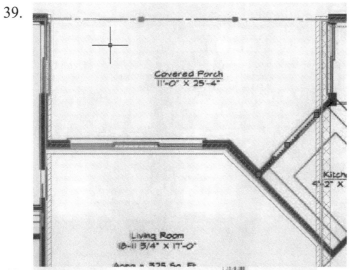

Place the west family room wall.

Verify that the distance from interior face to interior face is 20′ 0″.

39.

Trace over the floor plan to place the walls for the covered porch.

40.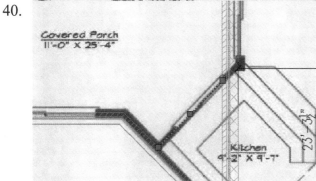

Draw a line at a 45° angle to designate the wall for the kitchen.

41. Locate the **Stud-3.5 Rigid 1.5 Air-1 Brick-4 Wall** style.
Right click and **Apply Tool Properties to → Linework** and select the angled line.

42.  Use the BREAK tool to divide the walls that need to be split into the two different styles.

43. 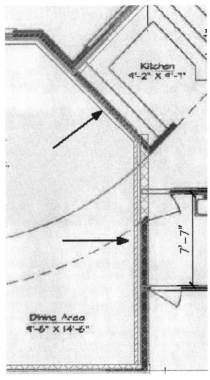 The walls indicated should be broken using the BREAK tool so one segment can remain exterior and one segment can be changed to the interior wall style.

44.

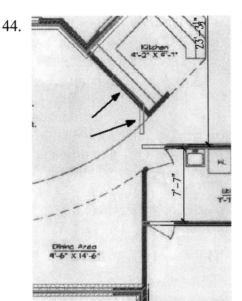

Change the interior wall segments to the interior wall style.

45.

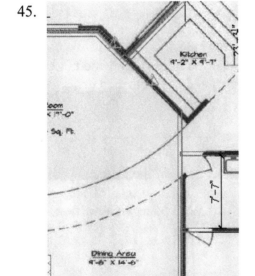

Use the EXTEND tool to extend the interior walls.

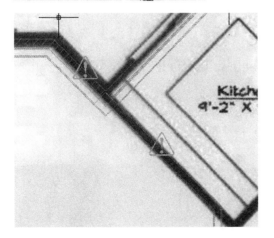

A triangle symbol with an exclamation point indicates that you have a wall interference condition – usually a wall on top of a wall.

46. 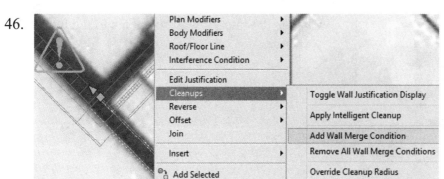 Select the interior wall with the interference condition. Right click and select **Cleanups → Add Wall Merge Condition**.

47. 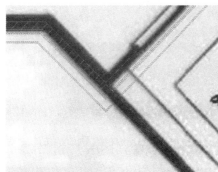 Select the two exterior walls where it is interfering.

The walls will merge and clean up the intersection area.

48. 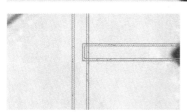 Zoom into the area near the utility room and notice some of the walls may need to be cleaned up as well.

49. 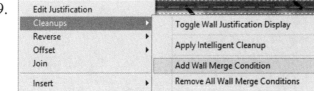 Select one of the interior walls. Right click and select **Cleanups → Add Wall Merge Condition**.

50. 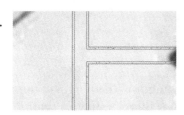 Select both walls.

The wall intersection cleans up.

51. 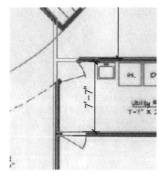 Repeat for the south utility wall.

52.

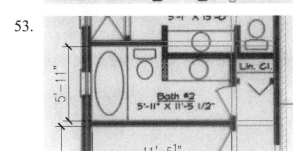

Zoom into the Bedroom #2 area.

Use FILLET to eliminate the extra interior walls. Select the walls at the locations indicated to clean up the room.

53.

Offset the south bathroom wall 6' 4.25".

Verify that the distance from interior face to interior face is 5' 11".

54.

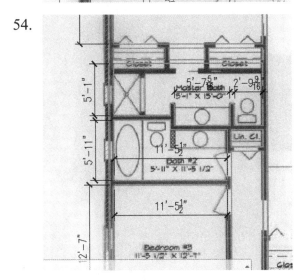

Add the interior walls for the lavatory areas.

Use the wall style **Stud-4 GWB-0.625 Each Side**.

55.

Unlock the image layer.

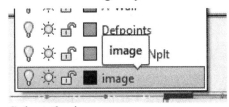

Select the image.
Right click and select Image→Adjust.

56. You can adjust how much of the image you see so it doesn't interfere with your work.

Alternatively, you can freeze the image layer or change the transparency of the layer.

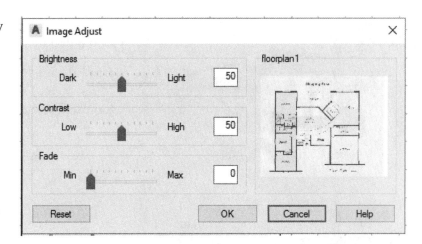

57. You should have a completed floor plan.

Save as *ex3-6.dwg*.

Exercise 3-7:

Adding Doors

Drawing Name: Ex3-6.dwg
Estimated Time: 45 minutes

This exercise reinforces the following skills:

❑ Adding Doors
❑ Door Properties

1. Open *ex3-6.dwg*.

2. Right click on the Command prompt and select **Options.**

3. Activate the **Profiles** tab.

Set the AutoCAD Architecture (US Imperial) profile as current.

4.    Thaw the image layer so you can see where doors are located if you froze that layer or adjust the image so you can see the door locations.

5.   Open the Design Tools palette.

6.   **Bifold - Double**   Locate the **Bifold-Double** door on the Doors tab on the Tools palette.

By changing the profile, more palettes are now available for you.

7.   Highlight the **Bifold - Double** door.
Right click and select **Properties**.

Apply Tool Properties to
Import 'Bifold - Double' Door Style
Door Styles...

Cut
Copy

Delete
Rename

Specify Image...
Refresh Image
Monochrome
Set Image from Selection...

Properties...

8.   Expand the **Dimensions** section.
Set the size to
4'-6" x 6'-8".

Set the Opening percent to **50**.

Image: Name:
Bifold - Double
Description:
Double bifold doors   More Info

Standard sizes   4'-6" X 6'-8"
A Width   --
B Height   --
   Measure to   --
   Opening per... 50

If you left click in the Standard sizes field, a down arrow will appear...select the down arrow and you will get a list of standard sizes. Then, select the size you want.

A 25% opening will show a door swing at a 45-degree angle.
The value of the Opening percentage determines the angle of the arc swing.
A 50% value indicates the door will appear half-open at a 90-degree angle.

9.

Location		▲
✳ Relative to grid	--	
✳ Position along wall	Offset/Center	⌄
✳ Automatic offset	6"	
✳ Justification	--	
Vertical alignment	--	
Head height	--	
Threshold height	--	

Expand the **Location** section.

Set Position along wall to **Offset/Center**. This will allow the user to snap to the center position along the wall.

Press **OK** to close the Properties dialog.

10. Place the Bifold - Double doors at the two closets.

The orientation of the door swing is determined by the wall side selected.

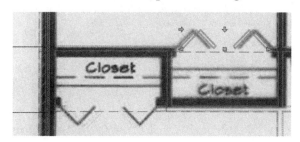

In both cases, you want to select the outside face of the wall.
Center the closet door on each wall.

11.

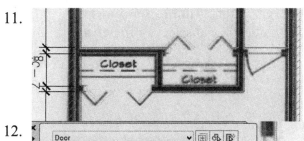

Place the **Bifold - Double** door at each of the closets located in Bedroom #2 and Bedroom #3.

12.

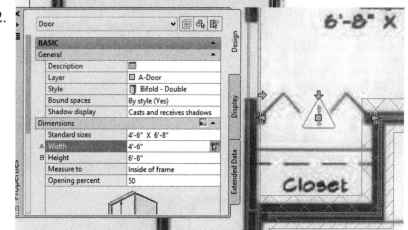

Place the **Bifold - Double** door at the closet next to the entry way.

The exclamation mark indicates that the door is too wide for the wall.

13.

Shadow display	Casts and receives shadows	
Dimensions		⌄ ▲
Standard sizes	4'-0" X 6'-8" (Custom Size)	
A Width	4'-0"	
B Height	6'-8"	
Measure to	Inside of frame	
Opening percent	50	

Select the door.

In the Properties palette, change the width of the Bifold - Double door to **4' 0"**.

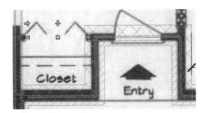

The door updates and the warning symbol disappears.

The door now fits.

14. 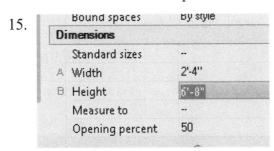 Locate the **Bifold - Single** door on the Doors tab of the Design Tools palette.

15. 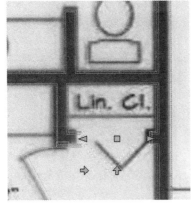 In the Properties palette:

Set the door to use the Standard Size **2' 4" x 6' 8"**.
Set the Opening percent to **50**.
Press **OK** to close the Properties palette.

Bound spaces	By style
Dimensions	
Standard sizes	--
A Width	2'-4"
B Height	6'-8"
Measure to	--
Opening percent	50

16. 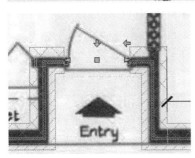 Place the door in the Linen Closet near the lavatories.

17.  Locate the **Hinged - Single - Exterior** door on the Doors tab of the Design Tools palette.

18.

A Width	3'-0"
B Height	6'-8"
Measure to	--
Swing angle	30

In the Properties palette, set the door to use the size **3' 0" x 6' 8"**.

Set the Swing angle to **30**.

19. Select the side of the wall that will be used for the door swing and place the entry door.

20. Locate the **Hinged - Single** door on the Doors tab of the Design Tools palette.

21.

A	Width	2'-6"
B	Height	6'-8"
	Measure to	--
	Swing angle	30

In the Properties palette, set the door to use the size **2′ 6″ x 6′ 8″**.

Set the Swing angle to **30**.

22.

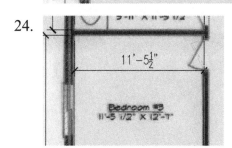

* Relative to grid --
* Position along wall Offset/Center
* Automatic offset 4 7/8"
* Justification --
* Vertical alignment --

Set the Position along wall to **Offset/Center**.

23.

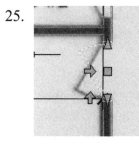

Place the door in Bedroom #2.

24.

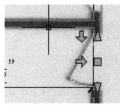

Place the door in Bedroom #3.

The swing is on the correct side but not the correct direction.

25.

Select the door so it highlights.

The horizontal arrow flips the orientation of the door to the other side of the wall.

The vertical arrow flips the orientation of the door swing.

Left click on the vertical arrow.

The door updates to match the floor plan image.

26. 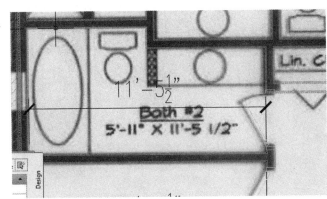 Place a **Hinged - Single** door in Bath #2.

27. 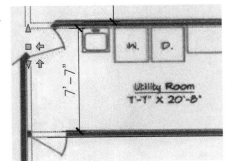 Place a **Hinged - Single** door in the Utility Room.

Set the swing angle to **70**.

28. 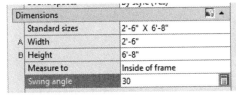 Locate the **Hinged - Single - Exterior** door on the Doors tab of the Design Tools palette.

29.

Dimensions	
Standard sizes	2'-6" X 6'-8"
A Width	2'-6"
B Height	6'-8"
Measure to	Inside of frame
Swing angle	30

In the Properties palette, set the door to use the Standard Size **2' 6" x 6' 8"**.

Set the Swing angle to **30**.

30. 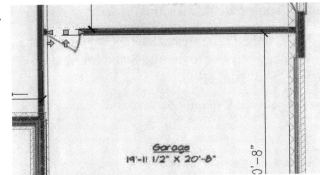 Place the door between the Utility Room and the Garage.

31. 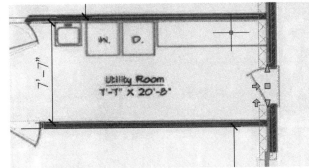 Place the door on the east wall of the Utility Room.

32. 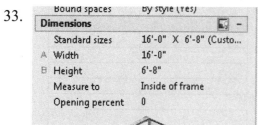 Locate the **Overhead - Sectional** door on the Doors tab of the Design Tools palette.

33. In the Properties palette, set the door to use the Size **16′ 0″ x 6′ 8″**.

Set the Opening percent to **0**.

34. Place the garage door.

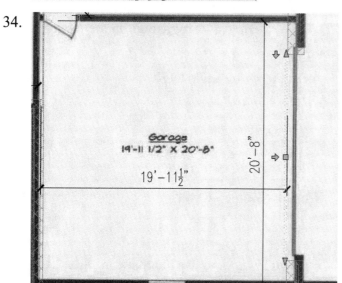

35. 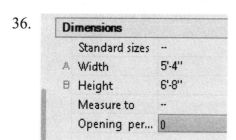 Locate the **Sliding - Double - Full Lite** door on the Doors tab of the Design Tools palette.

36. In the Properties palette, set the door to use the Standard Size **5′ 4″ x 6′ 8″**.

Set the Opening percent to **0**.

37. Place the door in the family room.

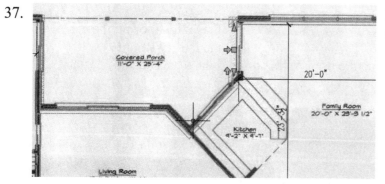

38. 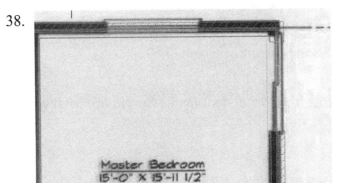 Place a second **Sliding - Double - Full Lite** door on the east wall of the Master Bedroom.

39.

	Standard sizes	10'-0" X 6'-8" (Custo...
A	Width	10'-0"
B	Height	6'-8"
	Measure to	Inside of frame
	Opening percent	50

Set the door to use the size **10' 0" x 6' 8"**.
Set the Opening Percent to **50**.

40. 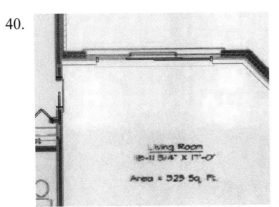 Center the door on the north wall of the Living Room.

41. 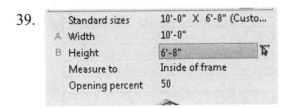 Locate the **Pocket - Single** door on the Doors tab of the Design Tools palette.

42.

Dimensions

	Standard sizes	--
A	Width	2'-6"
B	Height	6'-8"
	Measure to	--
	Opening per...	50

In the Properties palette, set the door to use the size: **2' 6" x 6' 8"**.

Set the Opening percent to **50**.

43. 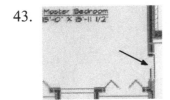 Place the door in the lower right corner of the Master Bedroom.

44.

Dimensions	
Standard sizes	2'-4" X 6'-8"
A Width	2'-4"
B Height	6'-8"
Measure to	Inside of frame
Opening percent	50

In the Properties palette,
set the door to use the Standard Size **2′ 4″ x 6′ 8″**.

Set the Opening percent to **50**.

45.

Center the pocket door on the lower horizontal wall between the Master Bedroom closets.

46.

Image layer is adjusted to be faded. Dimensions were moved to a layer named A-Anno-Dim and then frozen.

This is the floor plan so far.

47. Save as *ex3-7.dwg*.

Switch to an isometric view and you will see that your model is 3D.

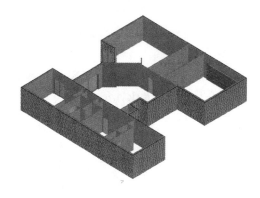

Look at the model using different visual styles. Which style do you like best? The model shown uses a Shaded visual style as defined by the dialog shown.

Exercise 3-8:
Create an Arched Opening Tool

Drawing Name: ex3-7.dwg
Estimated Time: 10 minutes

This exercise reinforces the following skills:

- ❑ Copying Tools
- ❑ Tool Properties

1. 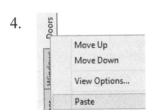 Open *ex3-7.dwg*.

2. ⌐ Cased Opening   Locate the **Cased Opening** tool on the Doors palette.

3. 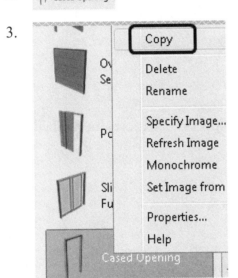   Right click and select **Copy**.

4. Select the **Doors** tab.
   Right click and select **Paste**.

5.

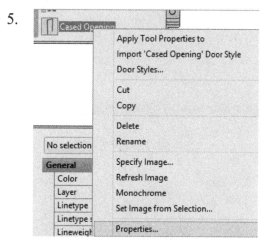

The copied tool is located at the bottom of the palette.

Highlight the copied tool.
Right click and select **Properties**.

6.

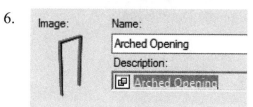

Change the Name to **Arched Opening**.
Change the Description to **Arched Opening**.
Press **OK**.

7.

Expand the General section.
Set the Description to **Creates an Arched Opening**.
Press **OK**.

8.

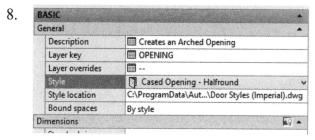

Set the Layer key to **OPENING**.
Set the Style to **Cased Opening-Half round**.

Press **OK**.

 The tool is defined in the palette.

9.   Save as *ex3-8.dwg*.

Exercise 3-9:
Adding an Opening

Drawing Name: ex3-8.dwg
Estimated Time: 15 minutes

This exercise reinforces the following skills:

- ❑ Adding Openings
- ❑ Opening Properties
- ❑ Copying Tools
- ❑ Set Image from Selection

Openings can be any size and elevation. They can be applied to a wall or be freestanding. The Add Opening Properties allow the user to either select a Pre-defined shape for the opening or use a custom shape.

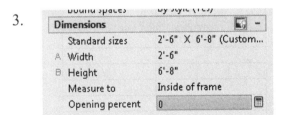

An opening will be added to the upper wall between the Master Bedroom closets.

1. Open *ex3-8.dwg*.

2. 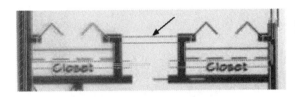 Select the **Arched Opening** tool.

3. | Dimensions | | |
    | --- | --- | --- |
    | Standard sizes | 2'-6" X 6'-8" (Custom... |
    | A Width | 2'-6" |
    | B Height | 6'-8" |
    | Measure to | Inside of frame |
    | Opening percent | 0 |

    In the Properties palette, set the door to use the size **2' 6" x 6' 8"**.

4. | Location | | - |
    | --- | --- | --- |
    | ✳ Relative to grid | No |
    | ✳ Position along wall | Offset/Center |
    | ✳ Automatic offset | 6" |
    | ✳ Justification | Center |

    Expand the Location section in the Properties palette.
    Set the Position along wall to **Offset/Center**.
    Set the Automatic offset to **6" [300.00]**.

5.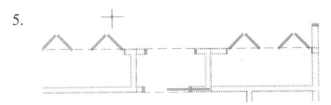

    Place the arched opening in the wall between the closets in the Master Bedroom. Center it on the wall.

6.

Dimensions	
Standard sizes	3'-0" X 6'-8"
A Width	3'-0"
B Height	6'-8"
Measure to	Inside of frame
Opening percent	50

In the Properties palette, set the door to use the size **3′ 0″ x 6′ 8″**.

7.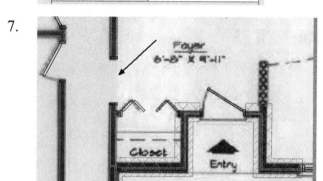

Place the Arched Opening on the left side of the Foyer above the Entry.

8. Use the View tools on the View ribbon **View → NE Isometric** and **3D orbit** to view the arched opening.

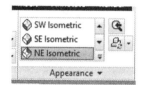

9. On the View ribbon,

Switch to a **Shades of Gray** display.

If your walls are reversed, you can change the orientation in the plan/top view.

10. Set the **Materials/Textures On**.

- Materials / Textures Off
- Materials On / Textures Off
- Materials / Textures On

11. Set to **Full Shadows**.

Note how the display changes.

- No Shadows
- Ground Shadows
- Full Shadows

When materials, textures, and shadows are enabled, more memory resources are used.

12. 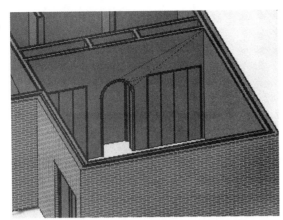 Locate the Arched Opening placed in the Master Bedroom.

13. 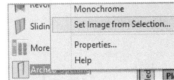 Select the **Arched Opening** icon on the tool palette. Right click and select **Set Image from Selection...** Pick the arched opening you created. Press **Enter**.

A dialog box allows you to choose which object to use for the image selection.

 If you have Selection Cycling enabled, you will see a selection dialog box.

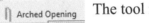 Select **Opening**.

Press **Enter**.

You can select more than one object for your image selection.

 The tool icon updates with the new image.

14. Select the Work tab to view your model.

15. Save the file as *ex3-9.dwg*.

Exercise 3-10:

Add Window Assemblies

Drawing Name: ex3-9.dwg
Estimated Time: 30 minutes

This exercise reinforces the following skills:

    ❑   Add Windows

1.   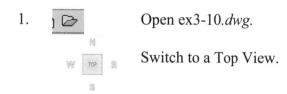   Open ex3-10.*dwg*.

          Switch to a Top View.

2.      Set the View style to **2D Wireframe**.

Remember you can change the view settings in the upper left corner of the display window.

3.   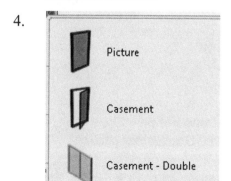   Activate the **Design Tools** from the Home ribbon, if they are not launched.

4.   Select the Windows tab of the Tool palette.

       Picture

       Casement

       Casement - Double

5.   Picture   Select the **Picture** window.

6.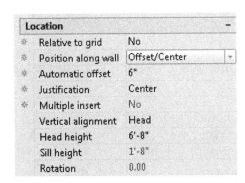

Expand the Dimensions section.
Set the size to **6'-0" x 5'-0"**.

7.

Expand the Location section.
Set the Position to **Offset/Center**.
Set the Automatic Offset to **6"**.

Location	−
* Relative to grid	No
* Position along wall	Offset/Center
* Automatic offset	6"
* Justification	Center
* Multiple insert	No
Vertical alignment	Head
Head height	6'-8"
Sill height	1'-8"
Rotation	0.00

8.

Select the midpoint of the north Master Bedroom wall.

9.

Standard sizes	9'-0" X 4'-0" (Custom...
A Width	9'-0"
B Height	4'-0"
Measure to	Outside of frame
Opening percent	0

On the Properties palette,
expand the Dimensions section.

Change the Width to **9'-0"**.
Change the Height to **4'-0"**.

10.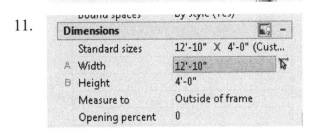

Place the window at the midpoint of the south wall in the Dining Area.

11.

Dimensions	−
Standard sizes	12'-10" X 4'-0" (Cust...
A Width	12'-10"
B Height	4'-0"
Measure to	Outside of frame
Opening percent	0

On the Properties palette,
expand the Dimensions section.

Change the Width to **12'-10"**.
Change the Height to **4'-0"**.

12.

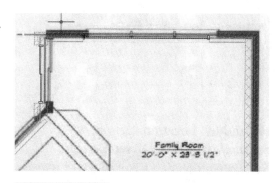

Place the window at the midpoint of the north wall for the Family Room.

13.

Casement - Double

Select the **Casement - Double** window on the Windows palette.

14.

Dimensions		
Standard sizes	4'-0" X 4'-0" (Custom...	
A Width	4'-0"	
B Height	4'-0"	
Measure to	Outside of frame	
Swing angle	0	

On the Properties palette,
expand the Dimensions section.

Change the Width to **4'-0"**.
Change the Height to **4'-0"**.

15.

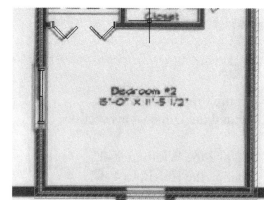

Place the window in the west wall of Bedroom #2.

16.

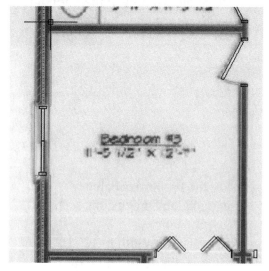

Place the window in the west wall of Bedroom #3.

17. Select the **Casement** window.

18. On the Properties palette, expand the Dimensions section.

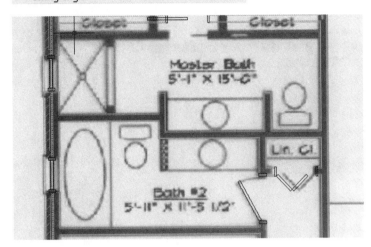

Change the Width to **2'-0"** .
Change the Height to **4'-0"**.

19. Place two windows on the west wall of the bathrooms.

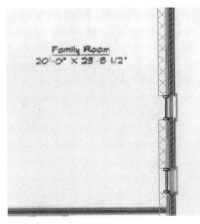

20. Place two windows on the east wall of the Family Room.

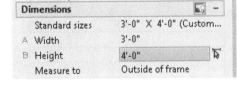

21. On the Properties palette, expand the Dimensions section.

Change the Width to **3'-0"** .
Change the Height to **4'-0"**.

22.

Place the window in the south wall of the Garage.

23. Save as *ex3-10.dwg*.

Lesson 4:
Space Planning

A residential structure is divided into three basic areas:

❑ Bedrooms: Used for sleeping and privacy
❑ Common Areas: Used for gathering and entertainment, such as family rooms and living rooms, and dining area
❑ Service Areas: Used to perform functions, such as the kitchen, laundry room, garage, and storage areas

When drawing your floor plan, you need to verify that enough space is provided to allow placement of items, such as beds, tables, entertainment equipment, cars, stoves, bathtubs, lavatories, etc.

AutoCAD Architecture comes with Design Content to allow designers to place furniture to test their space.

Exercise 4-1:
Creating AEC Content

Drawing Name: New
Estimated Time: 20 minutes

This lesson reinforces the following skills:

❑ Design Center
❑ AEC Content
❑ Customization

1. Start a new drawing using **QNEW** or press the + tab.

2. Activate the Insert ribbon.

Select **Attach**.

3. Browse for *ex3-10.dwg* and press **Open**.

File name: ex3-10
Files of type: Drawing (*.dwg)

4.  Accept the defaults.

Uncheck **Specify On-screen** for the insertion point.

Press **OK**.

5. 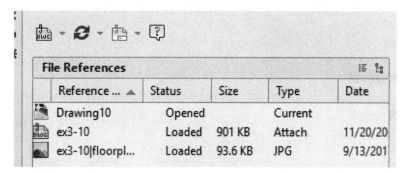 Launch the XREF Manager by selecting the small arrow in the Reference drop-down.

6. The palette lists the file references now loaded in the drawing. *Notice that the floorplan image file is listed as a reference.*

Reference ... ▲	Status	Size	Type	Date
Drawing10	Opened		Current	
ex3-10	Loaded	901 KB	Attach	11/20/20
ex3-10|floorpl...	Loaded	93.6 KB	JPG	9/13/201

7. 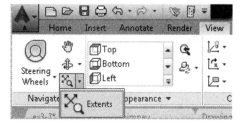 Zoom Extents.

You can do this by typing Z,E on the command line or using the tool on the View ribbon.

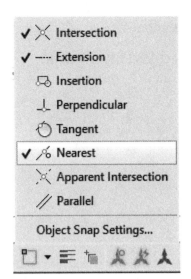

Enable the Nearest OSNAP to make it easier to place furniture.

8.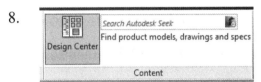

Launch the Design Center from the Insert ribbon or press Ctrl+2 or type **DC**.

9.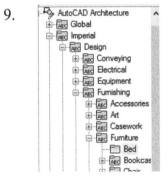

Select the AEC Content tab.

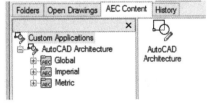

Browse to *AutoCAD Architecture/Imperial/Design/ Furnishing/ Furniture/Bed*.

10.

Select the *Double.dwg* file.
Right click and select **Insert**.

11.

Set Specify rotation to **Yes** in the Properties dialog.

12.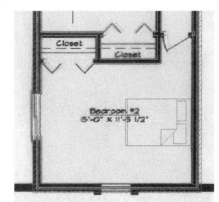

Place the bed in Bedroom #2.

Right click and select ENTER to exit the command.

13.

Twin

Locate the *Twin.dwg.*

14.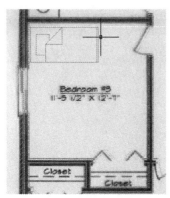

Place the *Twin.dwg* in Bedroom#3.

Right click and select ENTER to exit the command.

15.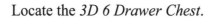

Browse to the *Bedroom* folder in the *Metric/Design/Domestic Furniture* folder.

16. Locate the *3D 6 Drawer Chest.*

3D 6 Drawer
Chest

17.

Place a *3D 6 Drawer Chest* in the Bedroom #2.
Right click and select ENTER to exit the
command.

18.

3D 3 Drawer
Chest

Add a 3D 3 Drawer Chest to Bedroom #3.

19.

Browse to the *Desk* folder under *AutoCAD
Architecture/Imperial/Design/Furnishing/Furniture/*.

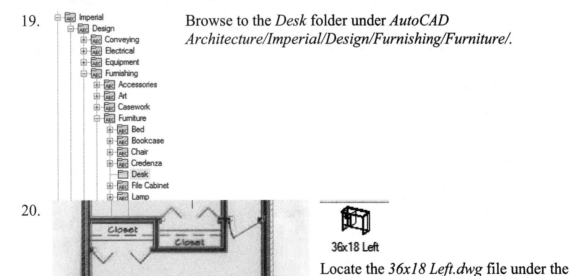

20.

36x18 Left

Locate the *36x18 Left.dwg* file under the
Desk folder.
Place into Bedroom #2.

21. 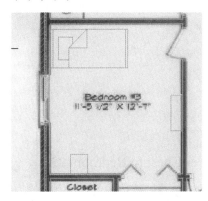 Locate the *6 Shelves.dwg* bookcase in the *Bookcase* folder.

22.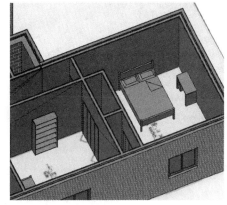

6 Shelves

Place in Bedroom #3.

23. Switch to a 3D view and use 3D Orbit to view the furniture placement.

24. Save as *ex4-1.dwg*.

Tips Tricks

The Space Planning process is not just to ensure that the rooms can hold the necessary equipment but also requires the drafter to think about plumbing, wiring, and HVAC requirements based on where and how items are placed.

As an additional exercise, place towel bars, soap dishes and other items in the bathrooms.

Exercise 4-2:
Furnishing the Common Areas

Drawing Name: Ex4-1.dwg
Estimated Time: 30 minutes

Common areas are the Living Room, Dining Room, and Family Room.

This lesson reinforces the following skills:

    □   Design Center
    □   Tool Palette
    □   Customization

You will need Internet access in order to complete this exercise. I have several blocks included to help you furnish your model. Feel free to explore and experiment with the different blocks.

1.   Open *ex4-1.dwg*.

2.   Launch the Design Center.

3.   Launch the Tool Palette

4.   Activate the **FF + E tab** on the tool palette. Scroll down to the Furnishings area.

5.   In the Design Center dialog, select the Folders tab.

Browse to the folder where you stored the files downloaded from the publisher.

6.   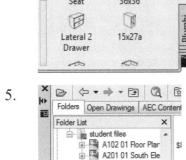 Drag and drop the file called **dining group** onto the FF +E palette.

7. Select the dining group tool.
Right click and select **Properties**.

8.  Change the name to **Dining Set**.

9. 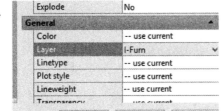 Set the Prompt for rotation to **Yes**.

10. Set Layer **to I-Furn**.

This ensures that the block will be placed on the I-Furn layer regardless of what layer is current.

11. Set the Color, Linetype, Plot Style, and Lineweight to ByLayer.

This means that the I-Furn layer properties will control the display of color, linetype, plot style and lineweight.

Press **OK** to close the Properties dialog.

12. 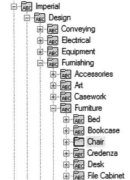 Activate the **AEC Content** tab on the Design Center.

Browse to *AutoCAD Architecture/Imperial/Design/Furnishing/ Furniture/Chair* folder.

13.  Drag and drop the **Lounge Circular** chair onto the Furniture palette.

14.  Select the **Lounge Circular** tool.

Right click and select **Properties**.

15.  Change the name to **Lounge Chair**.

Press **OK** to close the Properties dialog.

16.  Browse to *AutoCAD Architecture/Imperial/Design/Furnishing/ Furniture/Sofa* folder.

Drag and drop the **Love Seat 2** onto the Furniture palette.

17. Select the **Love Seat 2** tool.
     Right click and select **Properties**.

18.  Change the Name to **Love Seat 6'**.

19.  Browse to *AutoCAD Architecture/Imperial/Design/Furnishing /Furniture/Table* folder.

Drag and drop the **Coffee Table** onto the Furniture palette.

20. 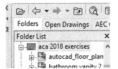 In the Design Center dialog, select the Folders tab.

Browse to the folder where you stored the files downloaded from the publisher.

21. 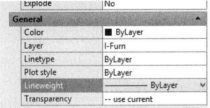 Drag and drop the file called entertainment center onto the palette.

22. Select the **entertainment center** tool. Right click and select **Properties**.

23. Change the name to **Entertainment Center**.

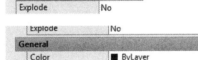

24. 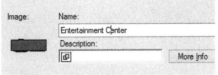 Set the Prompt for rotation to **Yes**.

25. Set Layer **to I-Furn**.

This ensures that the block will be placed on the I-Furn layer regardless of what layer is current. The available layers are in the active drawing.

26. Set the Color, Linetype, Plot Style, and Lineweight to ByLayer.

This means that the I-Furn layer properties will control the display of color, linetype, plot style and lineweight.

Press **OK** to close the Properties dialog.

27.

This is what your Furniture tool palette should look like.

There are additional furniture blocks in the student files you can add to the palette if you like.

28. Close the Design Center.

29.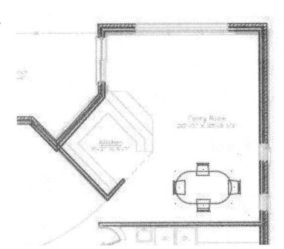

Add the **Dining Set** to the dining room area by the kitchen.

30.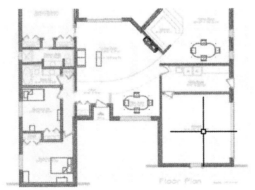

Furnish the living room.

31. Add a sofa and coffee table to the
family room.

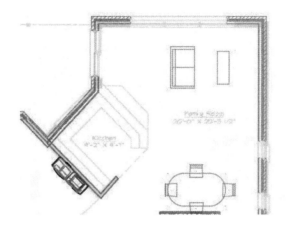

32. Save the file as *ex4-2.dwg*.

Tips & Tricks

If the drawing preview in the Design Center is displayed as 2D, then it is a 2D drawing.

Exercise 4-3:

Adding to the Service Areas

Drawing Name: Ex4-2.dwg
Estimated Time: 30 minutes

1. Open *ex4-2.dwg*.
   Select the Model tab.
   Switch to a top view.

2. Launch the Design Center.

3. 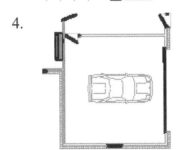 Select the AEC Content tab.
   Browse to the *AutoCAD Architecture/Imperial/Design/Site/ Basic Site/Vehicles* folder.

4.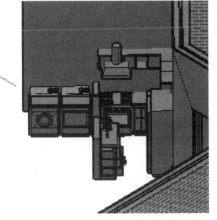

   Sub-Compact

   Drag and drop the *Sub-Compact* car into the garage.

   All that remains are the kitchen and laundry areas.

5. Use the Design Center to add casework, a kitchen sink, and an oven into the kitchen area.

   There are also models available to you in the exercise files. The layout shown is the kitchen block included in the exercise files scaled to fit into the space.

6. Our completed floor plan.

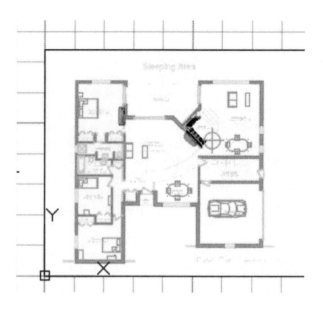

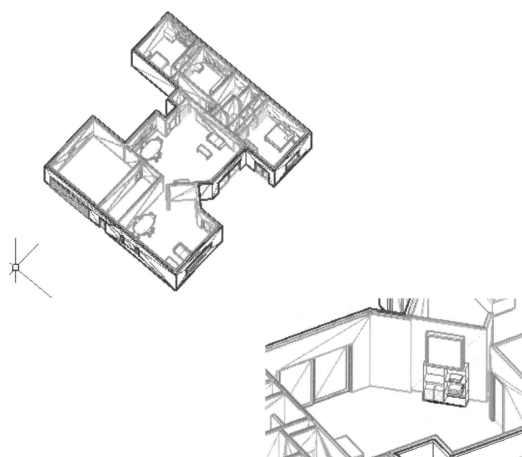

7. Save the file as *ex4-3.dwg*.

QUIZ 2

True or False

1. Once a door or window is placed, it cannot be moved or modified.

2. Openings can be any size and any elevation.

3. The Offset value when placing a door/window/opening determines how far the door/window/opening is placed from a selected point.

4. Door, window and opening dimensions can be applied using the Design Center.

Multiple Choice

5. Select the entity type that can NOT be converted to a wall:

   A. Line
   B. Polyline
   C. Circle
   D. Spline

6. A HOT grip is indicated by this color:

   A. GREEN
   B. BLUE
   C. RED
   D. YELLOW

7. To assign an image to a tool on the tool palette using existing geometry, use

   A. Assign image
   B. Set Image from selection
   C. Insert
   D. Import

ANSWERS:

1) F; 2) T 3) T; 4) F; 5) C; 6) C; 7) A

Q2-2

Notes:

Lesson 5:
Roofs

Roofs can be created with single or double slopes, with or without gable ends, and with or without overhangs. Once you input all your roof settings, you simply pick the points around the perimeter of the building to define your roof outline. If you make an error, you can easily modify or redefine your roof.

You need to pick three points before the roof will begin to preview in your graphics window. There is no limit to the number of points to select to define the perimeter.

To create a gable roof, uncheck the gable box in the Roof dialog. Pick the two end points for the sloped portion of the roof. Turn on the Gable box. Pick the end point for the gable side. Turn off the Gable box. Pick the end point for the sloped side. Turn the Gable box on. Pick the viewport and then press ENTER. A Gable cannot be defined with more than three consecutive edges.

Roofs can be created using two methods: ROOFADD places a roof based on points selected or ROOFCONVERT which converts a closed polyline or closed walls to develop a roof.

- ➢ If you opt to use ROOFCONVERT and use existing closed walls, be sure that the walls are intersecting properly. If your walls are not properly cleaned up with each other, the roof conversion is unpredictable.
- ➢ The Plate Height of a roof should be set equal to the Wall Height.
- ➢ You can create a gable on a roof by gripping any ridgeline point and stretching it past the roof edge. You cannot make a gable into a hip using grips.

Shape – Select the Shape option on the command line by typing 'S'.	Single Slope – Extends a roof plane at an angle from the Plate Height.	 end elevation view
	Double Slope – Includes a single slope and adds another slope, which begins at the intersection of the first slope and the height specified for the first slope.	 end elevation view
Gable – Select the Gable option on the command line by typing 'G'.	If this is enabled, turns off the slope of the roof place. To create a gable edge, select Gable prior to identifying the first corner of the gable end. Turn off gable to continue to create the roof.	 gable roof end
Plate Height – Set the Plate Height on the command line by typing 'PH'.	Specify the top plate from which the roof plane is projected. The height is relative to the XY plane with a Z coordinate of 0.	
Rise – Set the Rise on the command line by typing 'PR'.	Sets the angle of the roof based on a run value of 12.	A rise value of 5 creates a 5/12 roof, which forms a slope angle of 22.62 degrees.
Slope – Set the Slope on the command line by typing 'PS'.	Angle of the roof rise from the horizontal.	If slope angles are entered, then the rise will automatically be calculated.

Upper Height – Set the Upper Height on the command line by typing 'UH'.	This is only available if a Double Slope roof is being created. This is the height where the second slope will start.	
Rise (upper) – Set the Upper Rise on the command line by typing 'UR'.	This is only available if a Double Slope roof is being created. This is the slope angle for the second slope.	A rise value of 5 creates a 5/12 roof, which forms a slope angle of 22.62 degrees.
Slope (upper) – Set the Upper Slope on the command line by typing 'US'.	This is only available if a Double Slope roof is being created. Defines the slope angle for the second slope.	If an upper rise value is set, this is automatically calculated.
Overhang – To enable on the command line, type 'O'. To set the value of the Overhang, type 'V'.	If enabled, extends the roofline down from the plate height by the value set.	

	The Floating Viewer opens a viewer window displaying a preview of the roof.
	The match button allows you to select an existing roof to match its properties.
	The properties button opens the Roof Properties dialog.
	The Undo button allows you to undo the last roof operation. You can step back as many operations as you like up to the start.
	Opens the Roof Help file.

Exercise 5-1:

Creating a Roof using Existing Walls

Drawing Name: ex3-5.dwg
Estimated Time: 10 minutes

This exercise reinforces the following skills:

- ❑ Roof
- ❑ Roof Properties
- ❑ Visual Styles

1.   Open *ex3-2.dwg*.

2.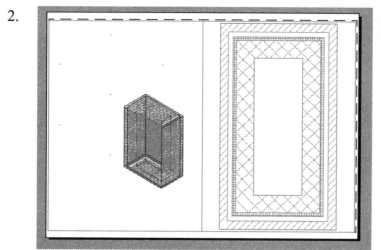

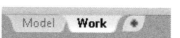

Select the Work icon.
Activate the right viewport that shows the top or plan view.

3. 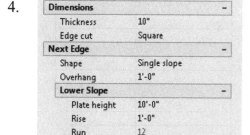 Select the **Roof** tool from the Home ribbon.

4.

Dimensions		
Thickness	10"	
Edge cut	Square	
Next Edge		
Shape	Single slope	
Overhang	1'-0"	
Lower Slope		
Plate height	10'-0"	
Rise	1'-0"	
Run	12	
Slope	45.00	

Expand the Dimensions section.
Set the Thickness to **10″ [254.00 mm]**.
Set the Shape to **Single slope**.
Set the Overhang to **1′-0″ [609.6 mm]**.

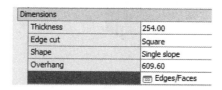

Dimensions	
Thickness	254.00
Edge cut	Square
Shape	Single slope
Overhang	609.60
	Edges/Faces

5.

Lower Slope	
Plate height	10'-0"
Rise	1'-0"
Run	12
Slope	45.00

Expand the Lower Slope section.
Set the Plate height to **10′-0″ [3000 mm]**.

The plate height determines the level where the roof rests.

6.

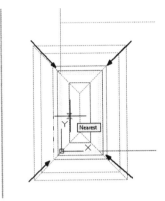

Lower Slope	
Plate height	3000.00
Rise	100.00
Run	100
Slope	45.00

Set the Rise to **1'-0" [100 mm]**.

7.

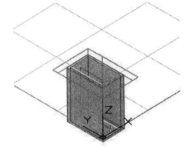

Pick the corners indicated to place the roof.

Press **Enter**.

8.

Activate the left 3D viewport.

9.

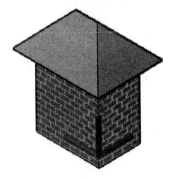

Activate the View ribbon.

Select the **Shades of Gray** under Visual Styles.

10.

Save as *ex5-1.dwg*.

In order to edit a roof you have to convert it to Roof Slabs.

Exercise 5-2:

Roof Slabs

Drawing Name: ex3-10.dwg
Estimated Time: 15 minutes

This exercise reinforces the following skills:

- ❑ Convert to Roof
- ❑ Roof Slab Tools

1. Open *ex3-10.dwg.*

2. Select the **Roof** tool from the Home ribbon.

3.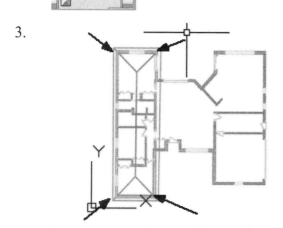
Select the four corners on the left wing to place a roof.

Press ENTER to complete the command.

4.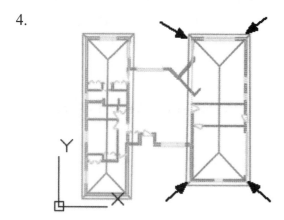
Repeat and select the four corners on the right wing to place a roof.

5.

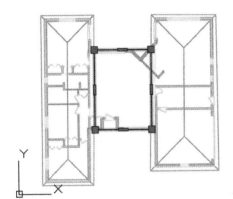

Draw a rectangle around the middle section of the floor plan using the rectangle tool.

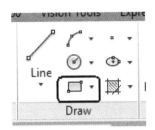

6.

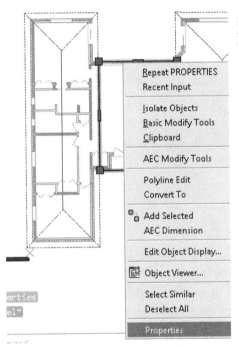

Select the rectangle.
Right click and select Properties.

7.

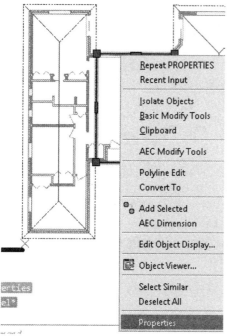

Change the elevation for the rectangle to 10'-0".

This puts it in-line with the top of the walls.

Material	ByLayer
Geometry	
Current Vertex	1
Vertex X	27'-8 3/8"
Vertex Y	51'-9 5/8"
Start segment width	0"
End segment width	0"
Global width	0"
Elevation	10'-0"

8.

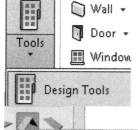

Launch the Design Tools from the Home ribbon.

9.

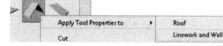

Locate the **Roof** tool on the Design palette. Right click and select **Apply Tool Properties to→Linework and Walls.**

10.

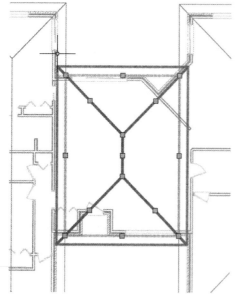

Select the rectangle.
Press ENTER to complete the selection.
When prompted to erase the linework, select **Yes**.
Press ENTER.
The middle roof is placed.

Press ESC to release any selection.

11. 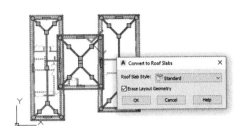 Select all three roofs.

The roof will highlight.

12. Select **Convert** from the Roof ribbon.

13. 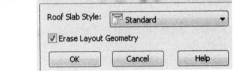 Enable the **Erase layout geometry** checkbox.

Press **OK**.

14. The roofs will change appearance slightly. They now consist of individual slabs.

15. Save as *ex5-2.dwg*.

Modify Roof Line

Drawing Name: ex5-2.dwg
Estimated Time: 15 minutes

This exercise reinforces the following skills:

- ❑   Modify Roof Slabs
- ❑   Modify Roof Line

1.   Open *ex5-2.dwg*.

2.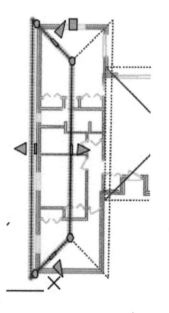

Erase the two roof slabs indicated.

Select the slabs, then press the **Delete** key on the keyboard.

3.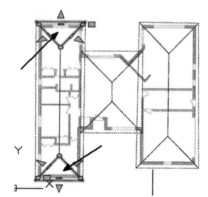

Select the left roof slab.

4.

Drag the point/grip at the bottom of the triangle up until it is horizontal to the upper point.

Look for the input that displays the angle at 180°.

5.   Repeat for the right side.

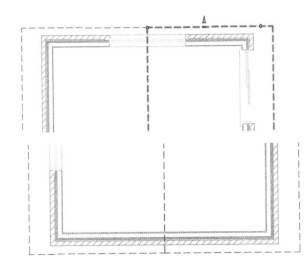

6.   Repeat for the bottom side.

If you are having problems creating a straight edge, draw a horizontal line and snap to the line. Then, erase the line.

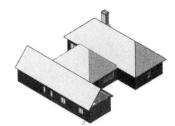

7.   Switch to a 3D view to see how the roof has changed.

View Style shown is Hidden, Materials On, Realistic Face Style.

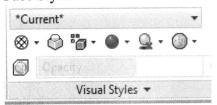

8.   Select the wall indicated.

9.   Select **Modify Roof Line** from the ribbon.

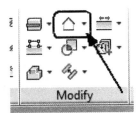

10. Type **A** for Autoproject.
Select the two roof slabs.
Press ENTER.

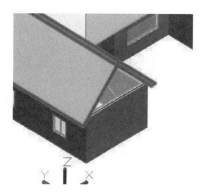

11. The wall will auto-project to the roof.

Press ESC to exit the command.

12.  Rotate the model so you can see the other side of the building.

13. 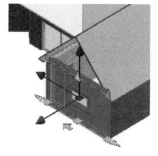 Select the wall indicated.

14. 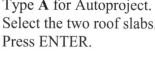 Select **Modify Roof Line** from the ribbon.

15. Type **A** for Autoproject.
Select the two roof slabs.
Press ENTER.

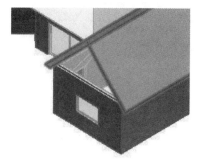

16. 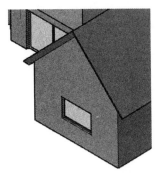 The wall will auto-project to the roof.

Press ESC to exit the command.

17. Save as *ex5-3.dwg*.

Exercise 5-4:
Trimming Roof Slabs

Drawing Name: ex5-3.dwg
Estimated Time: 30 minutes

This exercise reinforces the following skills:

- Roof Slabs
- Trim
- Grips

1. Open *ex5-3.dwg*.

2. Activate the **View** ribbon.

Select the **Top** view.

You can also use the control at the top left corner of the display window to change the view orientation.

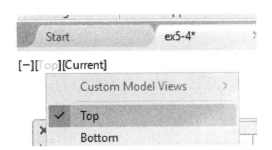

3.

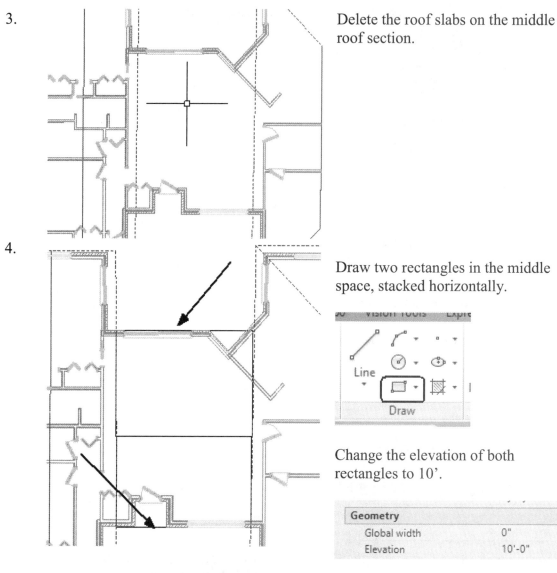

Delete the roof slabs on the middle roof section.

4.

Draw two rectangles in the middle space, stacked horizontally.

Change the elevation of both rectangles to 10'.

Geometry	
Global width	0"
Elevation	10'-0"

5.

Locate the **Roof Slab** tool on the Design Palette.
Right click and select **Apply Tool Properties to→Linework, Walls and Roof.**

6.

ROOFSLABTOOLTOLINEWORK Erase layout geometry? [Yes No] <No>: Y

Select both rectangles.
Press **ENTER**.
When prompted to erase layout geometry, type **Y** for yes.

7.

ROOFSLABTOOLTOLINEWORK Creation mode [Direct Projected] <Projected>:

Press **ENTER** to accept Projected.

8.

ROOFSLABTOOLTOLINEWORK Specify base height <8'-0">: 10'0"

Type **10' 0"** to set the base height to 10' 0".

9.

ROOFSLABTOOLTOLINEWORK Specify slab justification [Top Center Bottom Slopeline] <Bottom>:

Set the slab justification to **Bottom**.

10. 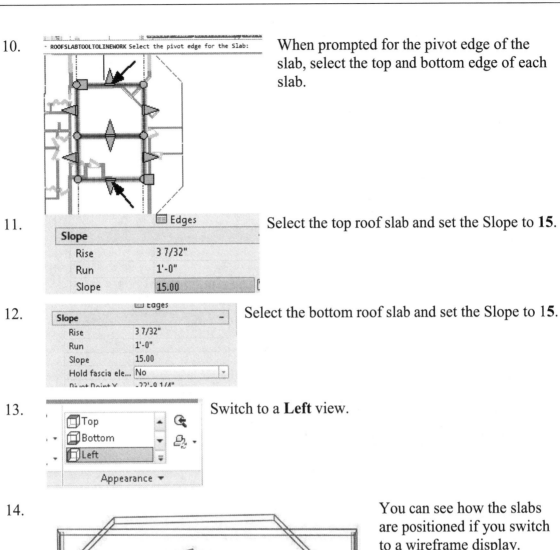 When prompted for the pivot edge of the slab, select the top and bottom edge of each slab.

11. Select the top roof slab and set the Slope to **15**.

Slope	
Rise	3 7/32"
Run	1'-0"
Slope	15.00

12. Select the bottom roof slab and set the Slope to **15**.

Slope	
Rise	3 7/32"
Run	1'-0"
Slope	15.00
Hold fascia ele...	No
Pivot Point X	-27'-9 1/4"

13. Switch to a **Left** view.

14. You can see how the slabs are positioned if you switch to a wireframe display.

15. 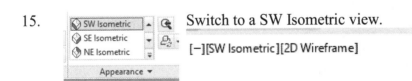 Switch to a SW Isometric view.

[–][SW Isometric][2D Wireframe]

16.

Select a roof slab.

17.

Select **Trim** on the ribbon.

18.

Select the other roof slab to use as the trimming object.

Then select the edge of the roof to be trimmed.

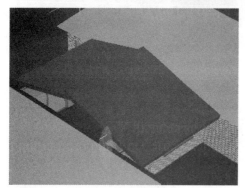

The roof slab adjusts.

19.

Activate the NE Isometric view.

20.

Select the untrimmed roof slab.

21.

Select **Trim** on the ribbon.

22.

Select the other roof slab to use as the trimming object.

Then select the edge of the roof to be trimmed.

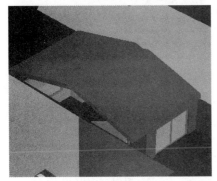

The roof slab adjusts.

23.

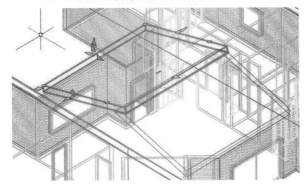

Select one of the roof slabs.

Use the grips to extend the slab into the adjoining roof slab.

24. 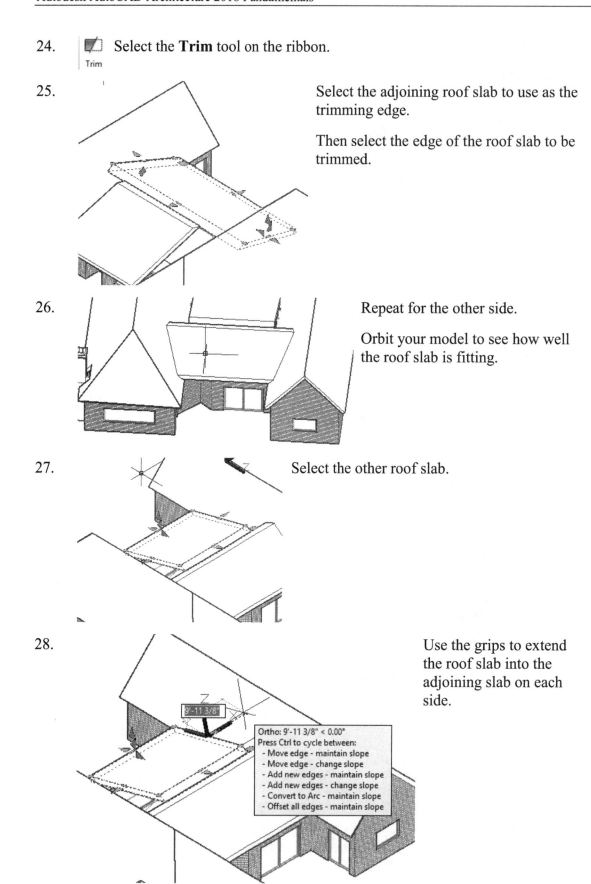 Select the **Trim** tool on the ribbon.

Trim

25. Select the adjoining roof slab to use as the trimming edge.

Then select the edge of the roof slab to be trimmed.

26. Repeat for the other side.

Orbit your model to see how well the roof slab is fitting.

27. Select the other roof slab.

28. Use the grips to extend the roof slab into the adjoining slab on each side.

Ortho: 9'-11 3/8" < 0.00°
Press Ctrl to cycle between:
- Move edge - maintain slope
- Move edge - change slope
- Add new edges - maintain slope
- Add new edges - change slope
- Convert to Arc - maintain slope
- Offset all edges - maintain slope

29.
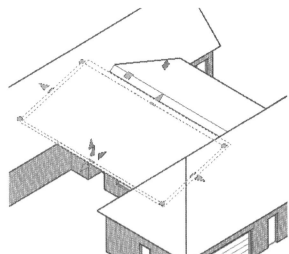
Select the TRIM tool on the ribbon.

30.
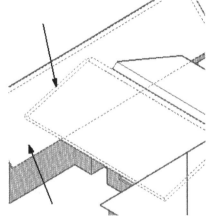
Select the roof slab to use as the trimming surface and the edge to be trimmed.

31.
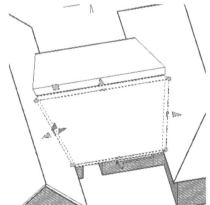
Repeat for the other side of the roof slab.

32.

Save as *ex5-4.dwg*.

Adding a Gable

Drawing Name: ex5-4.dwg
Estimated Time: 30 minutes

This exercise reinforces the following skills:

- Roof Slabs
- Add Hole
- Design Palette

1.   Open *ex5-4.dwg*.

2.   [−][Top][Shaded]   Switch to a **Top** view.

3.

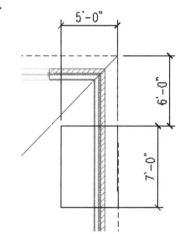

Draw a rectangle 6' 0" below the top edge of the far right roof slab.

The rectangle should be 5' 0" wide and 7' 0" high.

Do not add dimensions...those are for reference only.

If you use lines to place, use the PE polyline edit command to create a closed polyline.

4.

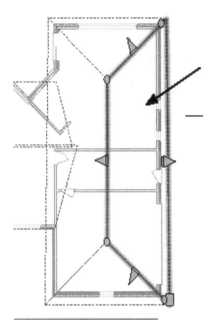

Select the far right roof slab.

5.

Select **Hole→Add** from the ribbon.

6. Select the rectangle.
Press **ENTER**.
Type **N** to erase the layout
geometry.

]- **ROOFSLABADDHOLE** Select closed polylines or connected solid objects to define holes:

7.

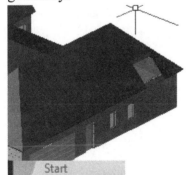

Switch to an isometric view and orbit around to see
the hole that was created.

8.

Switch back to a Top View.

9. 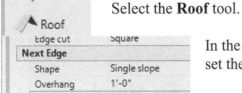 Select the **Roof** tool.

10. In the Properties pane,
set the Shape to **Single slope**.

11.

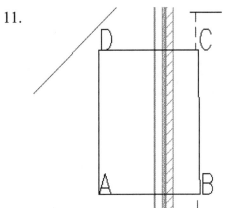

Select the two vertex points on the left side.

Select Point A and then Point B.

12.

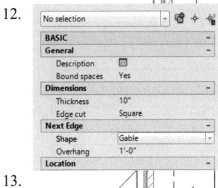

In the Properties pane,
set the Shape to **Gable**.

13.

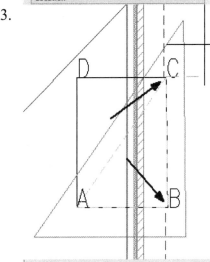

Select the two vertex points of the rectangle on the right side.

Select Point B and then Point C.

14.

Edge cut	Square
Next Edge	
Shape	Single slope
Overhang	1'-0"

Set the Shape to **Single slope.**

15.

Select Point D and then Point C.

The gable roof is created.

Press **ENTER**.

16. [—][Right][Shaded]

Switch to a Right view.

17.

You see the gable that was created.

Select the gable.

18.

Lower Slope	
Plate height	14' 0"
Rise	1'-0"

Set the Plate Height to **14' 0"**.

This will raise the gable.

19.

You can orbit the view slightly to check the gable position.

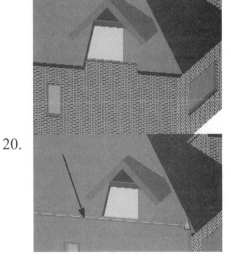

20.

Select the lower wall below the gable.

21.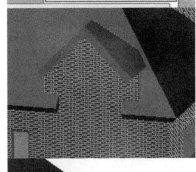

Select **Modify Roof Line** from the Ribbon.

22.

Select Auto project.

Select the roof gable.

The wall will adjust.

23.

Add a short exterior wall on each side of the gable.

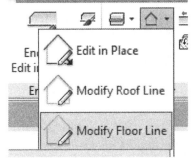

Use Modify Roof Line and Modify Floor line to trim the wall using the gable roof and roof slabs.

24. Save as ex5-5.rvt.

Exercise 5-6:

Roof Styles

Drawing Name: ex5-5.dwg
Estimated Time: 30 minutes

This exercise reinforces the following skills:

- ❑ Roof Styles
- ❑ Content Browser
- ❑ Design Palette

1.      Open *ex5-5.dwg*.

2.   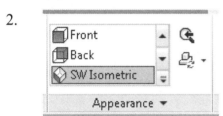   Switch to an SW Isometric view.

3.   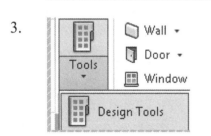   Launch the Design Tools.

4.   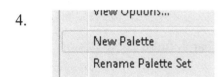   Right click on the Tool Palette and select **New Palette**.

5.      Rename the new palette – **Roofs**.

6.

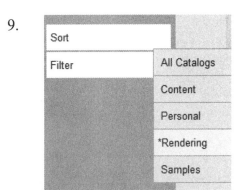

Launch the Content Browser.

7.

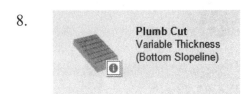

Type **plumb cut** in the search field.
Press **GO**.

8.

Plumb Cut
Variable Thickness
(Bottom Slopeline)

Drag and drop the **Plumb Cut Variable Thickness** onto the palette.

9.

Sort

Filter | All Catalogs
| Content
| Personal
| *Rendering
| Samples

Set the Filter on the Content Browser to **Rendering**.

10.

Type **asphalt** in the search field.
Press **GO**.

11.

Locate the rendering material for the Asphalt material.
Drag and drop it into the Roofs palette.

Minimize the Content Browser.

12. Activate the Manage ribbon.

Select the **Style Manager**.

13. Locate the Roof Slab Styles under Architectural Objects.

Note that there is only one slab style available in the drawing.

14. Right click on the Roof Slab Style and select **New**.

15. Rename the style **Asphalt Shingles**.

16.
Define three components:
- Asphalt Shingles with a thickness of ½" and Thickness Offset of 0.
- Membrane with a Thickness of 1/8" and Thickness offset of ½".
- Plywood Sheathing with a thickness of ¼" and Thickness offset of 5/8".

Use the preview to check the placement of each component.

17.  Create new material definitions for the three new component types:

- Asphalt Shingles
- Membrane
- Plywood Sheathing

18.  Highlight the Asphalt Shingles component.
Select **Edit Material Definition**.

19. 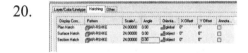 Select the General Medium Detail style override.

Set the Hatch to use **AR-RSHKE**.

20. 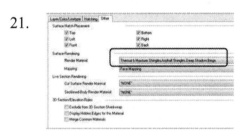 Set the Angles to 0.

21.  Select the **Other** tab.
Under Surface Rendering,
set the Render Material to the Asphalt Material that was placed in the file earlier.

Press **OK**.
Close the Style Manager.

22. 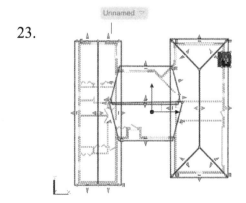 Switch to a top view.

23. Select all the roof slabs.

Window around the entire building.

Right click and select **Properties**.
Select **Roof Slab** from the drop-down list.

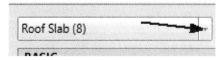

24.

On the Properties palette,
set the Style to **Asphalt Shingles**.

Press ESC to release the selection.

25.

Save as *ex5-6.dwg*.

Notes:

Lesson 6:
Structural Members

Architectural documentation for residential construction will always include plans (top views) of each floor of a building, showing features and structural information of the floor platform itself. Walls are located on the plan but not shown in structural detail.

Wall sections and details are used to show:

- the elements within walls (exterior siding, sheathing, block, brick or wood studs, insulation, air cavities, interior sheathing, trim)
- how walls relate to floors, ceilings, roofs, and eaves
- openings within the walls (doors/windows with their associated sills and headers)
- how walls relate to openings in floors (stairs).

Stick-framed (stud) walls usually have their framing patterns determined by the carpenters on site. Once window and door openings are located on the plan and stud spacing is specified by the designer (or the local building code), the specific arrangement of vertical members is usually left to the fabricators and not drafted, except where specific structural details require explanation.

The structural members in framed floors that have to hold themselves and/or other walls and floors up are usually drafted as framing plans. Designers must specify the size and spacing of joists or trusses, beams and columns. Plans show the orientation and relation of members, locate openings through the floor and show support information for openings and other specific conditions.

In the next exercises, we shall create a floor framing plan. Since the ground floor of our one-story lesson house has already been defined as a concrete slab, we'll assume that the ground level slopes down at the rear of the house and create a wood deck at the sliding door to the family room. The deck will need a railing for safety.

Autodesk AutoCAD Architecture includes a Structural Member Catalog that allows you to easily access industry-standard structural shapes. To create most standard column, brace, and beam styles, you can access the Structural Member Catalog, select a structural member shape, and create a style that contains the shape that you selected. The shape, similar to an AEC profile, is a 2D cross-section of a structural member. When you create a structural member with a style that you created from the Structural Member Catalog, you define the path to extrude the shape along.

You can create your own structural shapes that you can add to existing structural members or use to create new structural members. The design rules in a structural member style allow you to add these custom shapes to a structural member, as well as create custom structural members from more than one shape.

All the columns, braces, and beams that you create are sub-types of a single Structural Member object type. The styles that you create for columns, braces, and beams have the same Structural Member Styles style type as well. When you change the display or style of a structural member, use the Structural Member object in the Display Manager and the Structural Member Styles style type in the Style Manager.

If you are operating with the AIA layering system as your current layer standard, when you create members or convert AutoCAD entities to structural members using the menu picks or toolbars, AutoCAD Architecture assigns the new members to layers: A-Cols, A-Cols-Brce or A-Beam, respectively. If Generic AutoCAD Architecture is your standard, the layers used are A_Columns, A_Beams and A_Braces. If your layer standard is Current Layer, new entities come in on the current layer, as in plain vanilla AutoCAD.

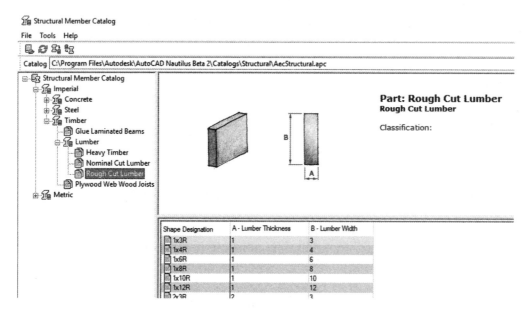

The Structural Member Catalog includes specifications for standard structural shapes. You can choose shapes from the Structural Member Catalog and generate styles for structural members that you create in your drawings.

> A style that contains the catalog shape that you selected is created. You can view the style in the Style Manager, create a new structural member from the style, or apply the style to an existing member.

> When you add a structural member to your drawing, the shape inside the style that you created defines the shape of the member. You define the length, justification, roll or rise, and start and end offsets of the structural member when you draw it.

You cannot use the following special characters in your style names:

- less-than and greater-than symbols (< >)
- forward slashes and backslashes (/ \)
- quotation marks (")
- colons (:)
- semicolons (;)
- question marks (?)
- commas (,)
- asterisks (*)
- vertical bars (|)
- equal signs (=)
- backquotes (`)

The left pane of the Structural Member Catalog contains a hierarchical tree view. Several industry standard catalogs are organized in the tree, first by imperial or metric units, and then by material.

	Open a catalog file - The default is the catalog that comes with ADT, but you can create your own custom catalog. The default catalog is located in the following directory path: \\Program Files\Autodesk AutoCAD Architecture R3\Catalogs\catalogs.
	Refresh Data
	Locate Catalog item based on an existing member – allows you to select a member in a drawing and then locates it in your catalog.
	Generate Member Style – allows you to create a style to be used.

We will be adding a wood framed deck, 4000 mm x 2750 mm, to the back of the house. For purposes of this exercise we will assume that the ground level is 300 mm below the slab at the back of the house and falls away so that grade level below the edge of the deck away from the house is 2 meters below floor level: - 2 m a.f.f. (above finish floor) in architectural notation. We will place the top of the deck floorboards even with the top of the floor slab.

We will use support and rim joists as in standard wood floor framing, and they will all be at the same level, rather than joists crossing a support beam below. In practice this means the use of metal hangers, which will not be drawn. Once the floor system is drawn we will add support columns at the outside rim joist and braces at the columns.

Exercise 6-1:
Creating Member Styles

Drawing Name: ex5-6.dwg
Estimated Time: 5 minutes

This exercise reinforces the following skills:

- ❑ Creating Member Styles
- ❑ Use of Structural Members tools

1.   Open *ex5-6.dwg*.

2.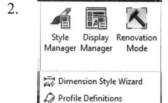
Activate the Manage ribbon.

Go to **Style & Display → Structural Member Catalog**.

3.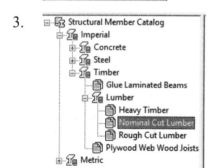
Browse to the **Imperial/Timber/Lumber/Nominal Cut Lumber** folder.

4.   In the lower right pane:

1x12	0.75	11.25
2x3	1.5	2.5
2x4	1.5	3.5
2x5	1.5	4.5

Locate the **2x4** shape designation.

5.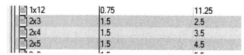
Right click and select **Generate Member Style**.

6. In the Structural Member Style dialog box, the name for your style automatically fills in.

Click **OK**.

7. In the lower right pane:

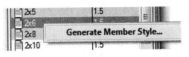

Locate the **2x6** shape designation.
Right click and select **Generate Member Style**.

8. In the Structural Member Style dialog box, the name for your style is displayed.

Click **OK**.

9. Close the dialog.
    Save as *ex6-1.dwg*.

Exercise 6-2:
Creating Member Shapes

Drawing Name: ex6-1.dwg
Estimated Time: 10 minutes

This exercise reinforces the following skills:

❑ Structural Member Wizard

1. Open *ex6-1.dwg*.
    Select the Work tab.
    Activate the right viewport.

2. Activate the **Manage** ribbon.

    Go to **Style & Display→Structural Member Wizard**.

    Style Manager   Display Manager   Renovation Mode

    ▤ Dimension Style Wizard
    ↻ Profile Definitions
    ⬚ Member Shape
    ↗ Insert as Polyline
    ⚙ Structural Member Catalog
    ⚙ Structural Member Wizard
    ▦ Show Display Overrides
    ◎ Style & Display

3.

    Select the **Cut Lumber** category under Wood.

    Press **Next**.

4.

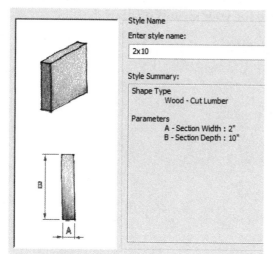

Set the Section Width to **2″**.

Set the Section Depth to **10″**.

Press **Next**.

5.

Enter the style name as **2x10**.

Press **Finish**.

6.

Activate the **Manage** ribbon.

Go to **Style & Display → Structural Member Wizard**.

7.

Select the **Cut Lumber** category under Wood.

Press **Next**.

8.

Set the Section Width to **4″**.

Set the Section Depth to **4″**.

Press **Next**.

9.

Enter the style name as **4x4**.

Press **Finish**.

Style Name
Enter style name:
4x4

Style Summary:
Shape Type
    Wood - Cut Lumber

Parameters
    A - Section Width : 4"
    B - Section Depth : 4"

10. Save as *ex6-2.dwg*.

Exercise 6-3:

Adding Structural Members

Drawing Name: ex6-2.dwg
Estimated Time: 25 minutes

This exercise reinforces the following skills:

- Creating Member Styles
- Use of Structural Members tools

1. Open *ex6-2.dwg*.

2. Activate the **Top** view.

Insert   Annotate   Render   View
Top
Bottom
Left
Appearance
World
Coordina

3. Freeze the **A-Roof-Slab** and **A-Roof** layers.

A-Roof
A-Roof-Slab

This will make it easier to place the members for the pergola.

4. Thaw the **image** layer.

G-Anno-Stnd-Scrn
image
image

5. Activate the Tools palette if it is not available.

To activate, launch from the Home ribbon.

Wall
Door
Window
Tools
Design Tools

6. 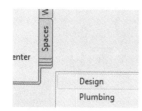 Activate the **Design** palette.
To do this, right click on the palette and select **Design**.

If you do not see the Design palette, check to see which profile you have active. Set the current profile to AutoCAD Architecture (US Imperial).

AutoCAD Architecture (US Imperial)
AutoCAD Architecture (US Imperial) <default>

We will place two columns at the locations indicated.

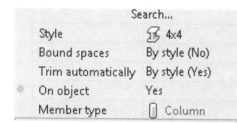

7. | Column   Select the **Column** tool on the **Design** tool palette.

8.
Search...	
Style	4x4
Bound spaces	By style (No)
Trim automatically	By style (Yes)
On object	Yes
Member type	Column

Set the Style to **4x4**.

9.
	Dimensions	
A	Start offset	0"
B	End offset	0"
C	Logical length	10'-0"
	Specify roll on screen	Yes
E	Roll	0.00
	Justify	Middle Center
	Justify cross-section	Maximum
	Justify components	Highest priority only

Set the Justify value to **Middle Center**.
Set the Logical Length to **10' -0"**.

10. 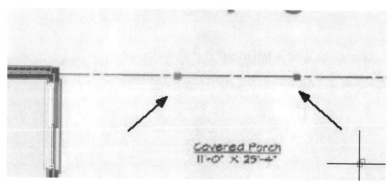 Place two columns at the locations indicated on the floor plan image.

11.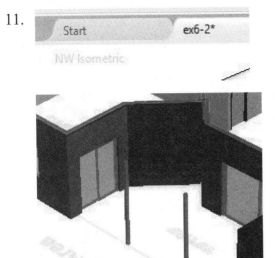

Switch to a **NW Isometric view**.

You should see two columns in the porch area.

12.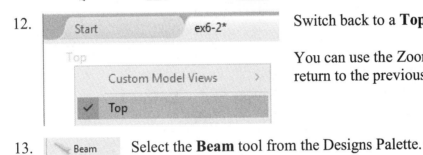

Switch back to a **Top** view.

You can use the Zoom Previous tool to return to the previous view.

13. Beam

Select the **Beam** tool from the Designs Palette.

14.

Style	2x4
Bound spaces	By style (No)
Trim automatically	By style (Yes)
On object	Yes
Member type	Beam
Dimensions	
Start offset	0"
End offset	0"
Logical length	1"
Roll	270.00
Layout type	Fill
Justify	Bottom Center
Justify cross-secti...	Maximum
Justify components	Highest priority only

Set the Style to **2x4**.

Set Justify to **Bottom Center**.

15.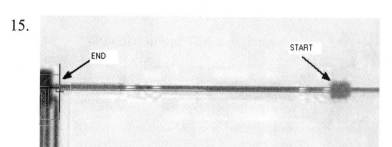

Select the node of the column as the start point and perpendicular point at the left wall as the end point.

16.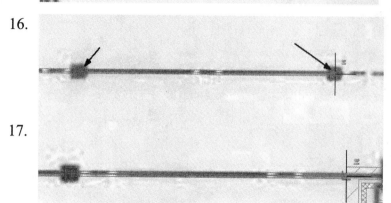

Place a second beam from the center of the left column to the center of the right column.

17.

Place a third beam starting at the node center of the column and perpendicular to the right wall.

18. Right click and select ENTER to exit the command.

19.

Select all three beams.

20.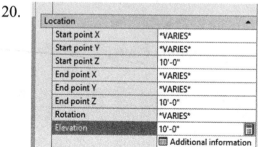

On the Properties palette:

Set the Start point Z to **10′ 0″**.
Set the End point Z to **10′ 0″**.

Note the Elevation updates to 10′ 0″.

Press ESC to release the selection set.

21. 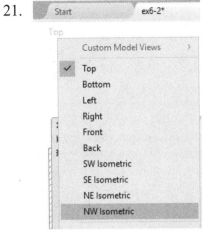 Activate the NW Isometric view.

22. 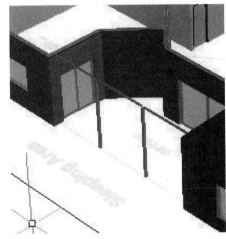 Inspect the columns and beams.

23. Save as *ex6-3.dwg*.

Notes:

QUIZ 3

True or False

1. Custom content can be located in any subdirectory and still function properly.

2. The sole purpose of the Space Planning process is to arrange furniture in a floor plan.

3. The Design Center only has 3D objects stored in the Content area because ADT is strictly a 3D software.

4. Appliances are automatically placed on the APPLIANCE layer.

5. When you place a wall cabinet, it is automatically placed at the specified height.

6. You can create tools on a tool palette by dragging and dropping the objects from the Design Center onto the palette.

Multiple Choice

7. A residential structure is divided into:

    A. Four basic areas
    B. Three basic areas
    C. Two basic areas
    D. One basic area

8. Kitchen cabinets are located in the _____ subfolder.

    A. Casework
    B. Cabinets
    C. Bookcases
    D. Furniture

9. Select the area type that is NOT part of a private residence:

    A. Bedrooms
    B. Common Areas
    C. Service Areas
    D. Public Areas

10. To set the layer properties of a tool on a tool palette:

    A. Use the Layer Manager
    B. Select the tool, right click and select Properties.
    C. Launch the Properties dialog
    D. All of the above

11. Vehicles placed from the Design Center are automatically placed on this layer:

    A. A-Site-Vhcl

    B. A-Vhcl

    C. C-Site Vhcl

    D. None of the above

12. The **Roof** tool is located on this tool palette:

    A. DESIGN

    B. GENERAL DRAFTING

    C. MASSING

    D. TOOLS

ANSWERS:

1) T; 2) F; 3) F; 4) F; 5) T; 6) T; 7) B; 8) A; 9) D; 10) B; 11) C; 12) A

Lesson 7:
Rendering & Materials

AutoCAD Architecture divides materials into TWO locations:

- Materials in Active Drawing
- Autodesk Library

You can also create one or more custom libraries for materials you want to use in more than one project. You can only edit materials in the drawing or in the custom library – Autodesk Materials are read-only.

Materials have two asset definitions –Appearance and Information. Materials do not have to be fully defined to be used. Each asset contributes to the definition of a single material.

You can define Appearance independently without assigning/associating it to a material. You can assign the same assets to different materials. Assets can be deleted from a material in a drawing or in a user library, but you cannot remove or delete assets from materials in locked material libraries, such as the Autodesk Materials library or any locked user library. You can delete assets from a user library – BUT be careful because that asset might be used somewhere!

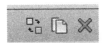

You can only duplicate, replace or delete an asset if it is in the drawing or in a user library. The Autodesk Asset Library is "read-only" and can only be used to check-out or create materials.

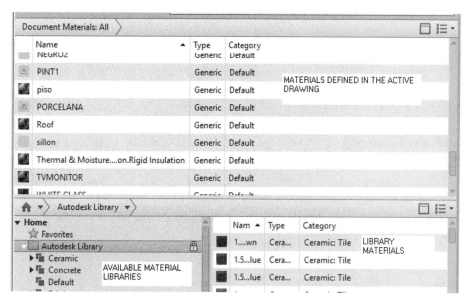

The Material Browser is divided into three distinct panels.

The Document Materials panel lists all the materials available in the drawing.

The Material Libraries list any libraries available.

The Library Materials lists the materials in the material libraries.

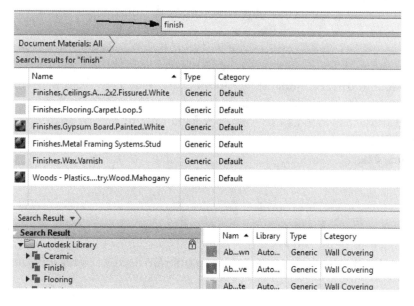

The search field allows you to search materials using keywords, description or comments.

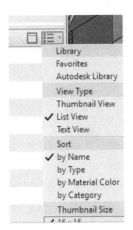

The View control button allows you to control how materials are sorted and displayed in the Material Libraries panel.

Material Editor

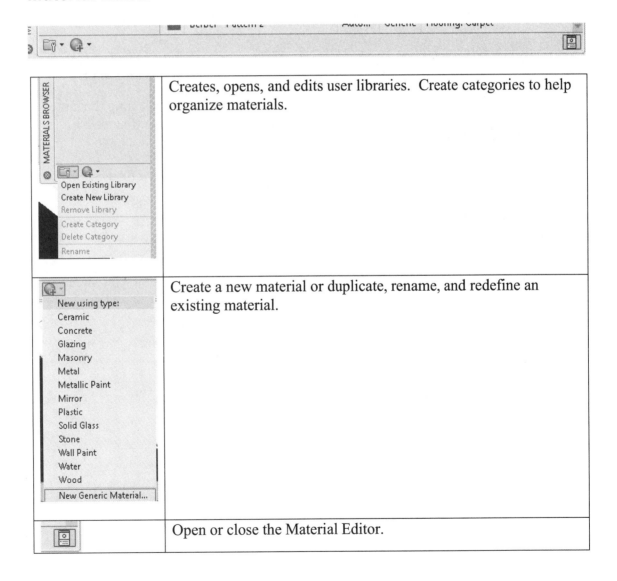

	Creates, opens, and edits user libraries. Create categories to help organize materials.
	Create a new material or duplicate, rename, and redefine an existing material.
	Open or close the Material Editor.

Exercise 7-1:
Modifying the Material Browser Interface

Drawing Name: ex6-3.dwg
Estimated Time: 10 minutes

This exercise reinforces the following skills:

    ❑   Navigating the Material Browser and Material Editor

1.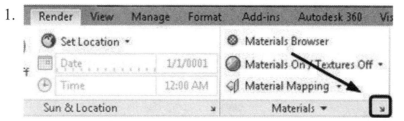

   Activate the **Render** ribbon.

   Select the small arrow in the lower right corner of the materials panel.

   This launches the Material Editor.

2.

   Launch the Material Browser by selecting the button on the bottom of the Material Editor dialog.

3.

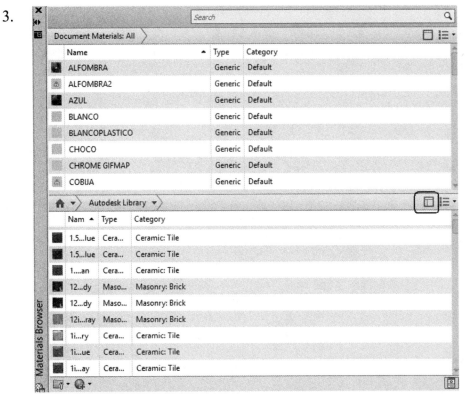

Make the Material Browser appear as shown.

Toggle the Materials Library pane OFF by selecting the top button on the Libraries pane.

4.

Toggle the Materials Library list OFF by selecting the top button on the Documents pane.

5.

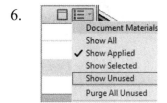

Toggle the panes ON by selecting the top button.

6. In the Document Materials panel, set the view to **Show Unused**.

Scroll through the list to see what materials are not being used in your drawing.

7. In the Material Library pane, change the view to sort **by Category**.

8. Close the Material dialogs.

9. Save as *ex7-1.dwg*.

The Appearance Tab

Appearance controls how material is displayed in a rendered view, Realistic view, or Ray Trace view.

A zero next to the hand means that only the active/ selected material uses the appearance definition.

Different material types can have different appearance parameters.

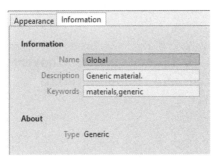

The Information Tab

The value in the Description field can be used in Material tags.

Using comments and keywords can be helpful for searches.

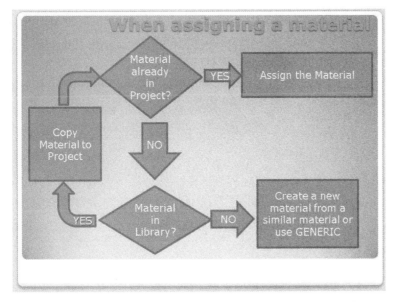

Use this flow chart as a guide to help you decide whether or not you need to create a new material in your project.

Exercise 7-2:
Copy a Material from a Library to a Drawing

Drawing Name: ex7.1.rvt
Estimated Time: 10 minutes

In order to use or assign a material, it must exist in the drawing. You can search for a material and, if one exists in the Autodesk Library, copy it to your active drawing.

This exercise reinforces the following skills:

- Copying materials from the Autodesk Materials Library
- Modifying materials in the drawing

1.   Go to the **Render** ribbon.

2.      Select the **Materials Browser** tool.

3.      If you see this dialog, select **Convert materials and add to Materials Browser**.

4.   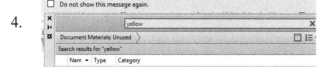   Type **yellow** in the search field.

There are no yellow materials in the drawing.

5.

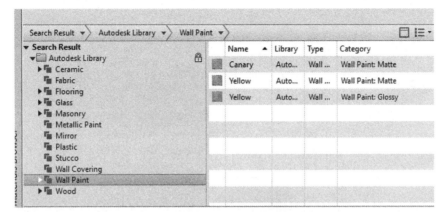

Highlight **Wall Paint** under Autodesk Library.

There are three yellow wall paints.

6.

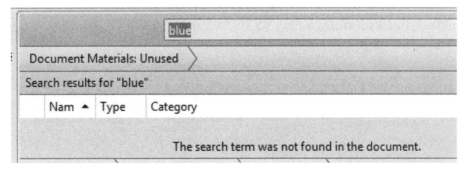

Locate the Yellow Matte Wall Paint.

Select the up arrow to copy the material up to the drawing.

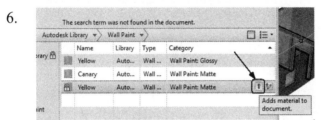

The yellow paint material is now listed in the Document Materials.

7.   Type **blue** in the search field.

	Nam ▲	Type	Category

blue

Document Materials: Unused

Search results for "blue"

The search term was not found in the document.

8.

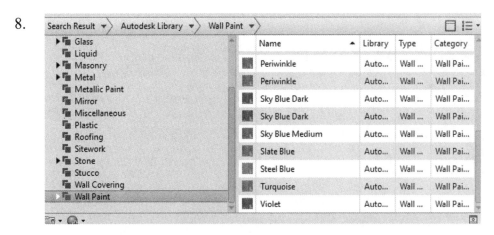

There are no blue materials in the drawing, but the Autodesk Library has several different blue wall paints available.

9.

Locate the **Periwinkle** Wall Paint and then copy it over to the drawing.

10.

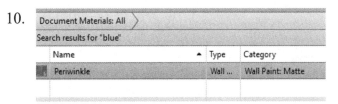

The blue wall paint is now available in the drawing.

11. Close the Materials dialog.

12. Save as *ex7-2.rvt*.

You have added material definitions to the active drawing ONLY.

Exercise 7-3:

Adding Color to Walls

Drawing Name: ex7-2.dwg
Estimated Time: 5 minutes

This exercise reinforces the following skills:

- Edit External Reference
- Materials
- Colors
- Display Manager
- Style Manager

1.   Open *ex7-2.dwg*.

2.   Zoom into the kitchen area. We are going to apply a different color to the kitchen wall.

     Select the wall indicated.

3.   Right click and select **Edit Wall Style**.

     Notice that there are two GWB components.
     One GWB is for one side of the wall and one is for the other.

4.   Rename them **GWB-Yellow** and **GWB-Blue**.

     To rename, just left click in the column and edit the text.

Index	Name	Priority	
1	GWB- Yellow	1210	5
2	GWB	1200	5
3	Stud	500	4
4	GWB	1200	5
5	GWB-Blue	1210	5

5.   Activate the Materials tab.

Component	Material Definition
GWB-Yellow	0Finishes.Plaster and Gypsum Board.Gypsum Wallboard.Painted.White
Stud	0Finishes.Metal Framing Systems.Stud
GWB-Blue	0Doors & Windows.Metal Doors & Frames.Steel Doors.Painted.White
Shrinkwrap	0Standard

Highlight the **GWB-Yellow** component.
Select the **New** button.

6.   Type **Finishes.Wall Paint.Yellow.**

     Press **OK**.

New Material

New Name: Finishes.Wall Paint.Yellow

OK   Cancel

7.
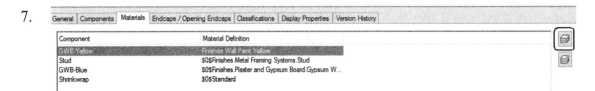

Highlight the **GWB-Yellow** component.
Select the **Edit** button.

8.
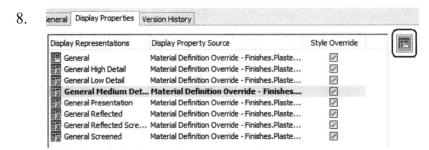

Activate the **Display Properties** tab.
Select the **Edit Display Properties** button.

9.
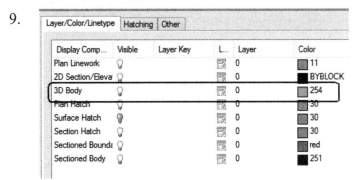

Activate the **Layer/Color/ Linetype** tab.

Highlight **3D Body**.

Left pick on the color square.

10.

Select the color **yellow**.

Press **OK**.

11.

Select the **Other** tab.

Select **Browse** next to Render Material.

12.
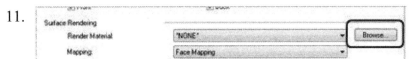

Scroll down to the Yellow color.

Highlight and press **OK**.

Note that the materials are in alphabetic order.

13.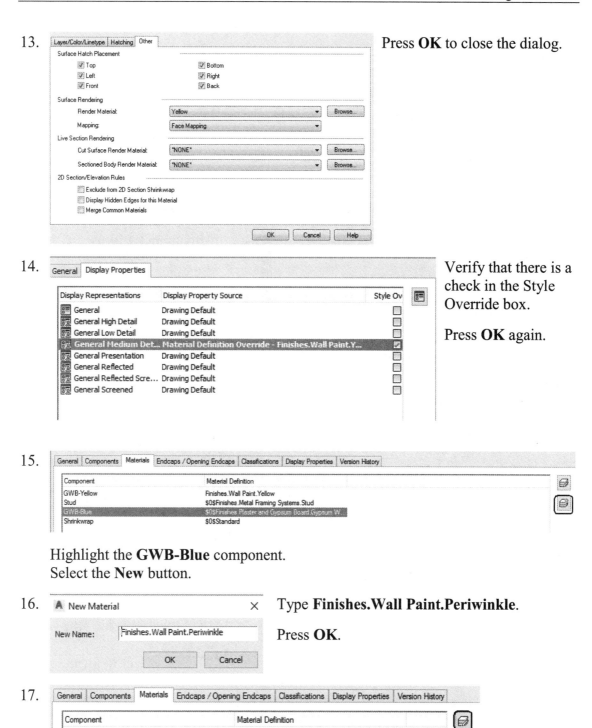

Press **OK** to close the dialog.

14. Verify that there is a check in the Style Override box.

Press **OK** again.

15. Highlight the **GWB-Blue** component.
Select the **New** button.

16. Type **Finishes.Wall Paint.Periwinkle**.

Press **OK**.

17. Highlight the **GWB-Blue** component.
Select the **Edit** button.

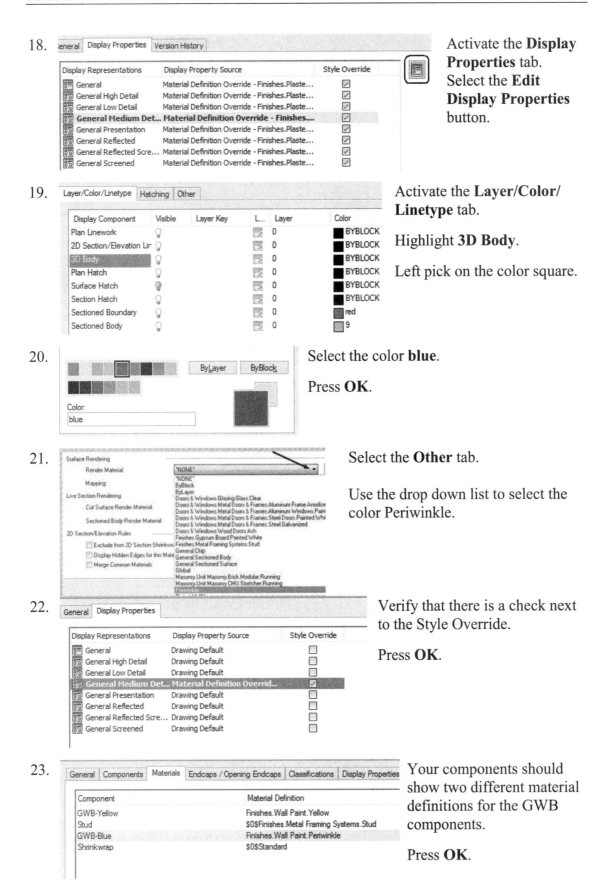

18. Activate the **Display Properties** tab. Select the **Edit Display Properties** button.

19. Activate the **Layer/Color/Linetype** tab.

Highlight **3D Body**.

Left pick on the color square.

20. Select the color **blue**.

Press **OK**.

21. Select the **Other** tab.

Use the drop down list to select the color Periwinkle.

22. Verify that there is a check next to the Style Override.

Press **OK**.

23. Your components should show two different material definitions for the GWB components.

Press **OK**.

24. 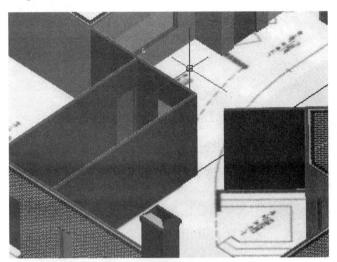 On the View ribbon,
change the display to **Shaded**.

25. Inspect the walls.

Notice that all walls using the wall style have blue on one side and yellow on the other.

In order to apply different colors to different walls, you would have to create a wall style for each color assignment.

26. To flip the color so walls are yellow/blue, simply use the direction arrows.

27. Save as *ex7-3.dwg*.

Exercise 7-4:
Reference Manager

Drawing Name: ex4-3.dwg
Estimated Time: 5 minutes

This exercise reinforces the following skills:

❑   Reference Manager

1.   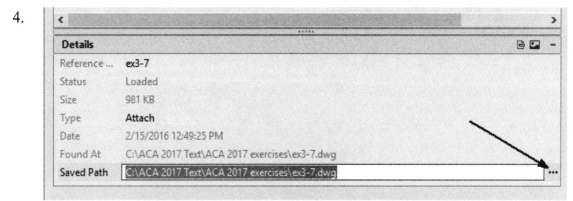   Open *ex4-3.dwg*.

2.   Activate the **Insert** ribbon.

Launch the Reference Manager by selecting the small arrow in the right corner of the Reference panel.

3.   Highlight the *ex3-10.dwg* file.

In the bottom pane, locate the Saved Path field.

4.

Details		
Reference ...	ex3-7	
Status	Loaded	
Size	981 KB	
Type	**Attach**	
Date	2/15/2016 12:49:25 PM	
Found At	C:\ACA 2017 Text\ACA 2017 exercises\ex3-7.dwg	
Saved Path	C:\ACA 2017 Text\ACA 2017 exercises\ex3-7.dwg	•••

Click on the … button to the far right of the saved path field.

5.

| File name: | ex7-3 | ⌄ | Open | ▾ |
| Files of type: | Drawing (*.dwg) | ⌄ | Cancel | |

Locate and select *ex7-3.dwg*.

Press **Open**.

6.

| Found At | C:\ACA 2017 Text\ACA 2017 exercises\ex7-3.dwg |
| Saved Path | C:\ACA 2017 Text\ACA 2017 exercises\ex7-3.dwg |

This changes the Saved Path to *ex7-3.dwg*.

This updates the external reference and replaces ex3-7 with ex7-3.

7.   Select the Refresh button.

The file updates. Notice the structural members that were added in the previous lesson. The colors on the interior walls should also display properly in Shaded display mode.

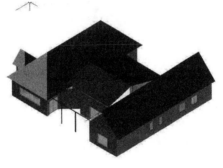

8.   Save the file as *ex7-4.dwg*.

Exercise 7-5:

Camera View

Drawing Name: ex7-4.dwg
Estimated Time: 5 minutes

This exercise reinforces the following skills:

- ❑ Camera
- ❑ Display Settings

1.   Open *ex7-4.dwg*.

2.   Use the View Cube to orient the floor plan to a Top View with North in correct position. *You can also use the right click menu on the upper left of the display window.*

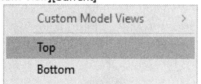

3.   Activate the Render ribbon.

   On the Camera panel,

   select **Create Camera**.

4. 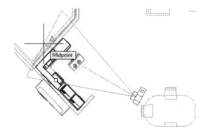   Place the Camera so it is in the dining area and looking toward the kitchen.

   Press ENTER.
   Select the camera so it highlights.

5.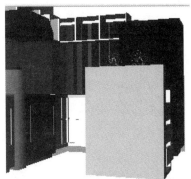

A Camera Preview window will open.

If you don't see the Camera Preview window, left click on the camera and the window should open.

Set the Visual Style to **Shaded**.

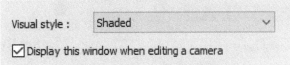

6.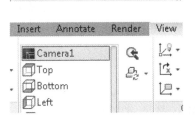

Right click on the camera and select **Properties**.

Change the Camera Z value to **3' 6"**.

This sets the camera at 3' 6" above floor level.

7.

Note that the Preview Window updates.

8.

Activate the View ribbon.

Notice the Camera1 view has been added to the view lists.

Left click to activate the Camera1 view.

9.

Change the Display Style to **Realistic**.

10.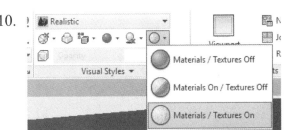

Set Materials/Textures **On**.

11. 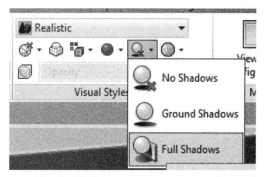 Set **Full Shadows** on.

12.  Set **Realistic Face Style** on.

 The Camera view should adjust as the different settings are applied.

13. Save as *ex7-5.dwg*.

In order to set location settings used in Rendering, you need to download and install a Hotfix from Autodesk's website. You must have administrative privileges on your Microsoft Windows® operating system to complete the installation process.

1.   Download the following executable on to your machine:
Autodesk_AutoCAD_2015_to_2018_Geolocation_Online_Maps_Hotfix.sfx.exe

2.   Close all software applications.

3.   Double click on the executable to extract the files. This action will also update all supported products on your machine with the Hotfix.

Exercise 7-6:
Create Rendering

Drawing Name: ex7-5.dwg
Estimated Time: 5 minutes

This exercise reinforces the following skills:

- ❏ Render
- ❏ Display Settings

1.   Open *ex7-5.dwg*.

2. 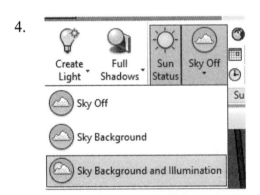 Set the Top view active.

3.   **Render**   Activate the Render ribbon.

4. 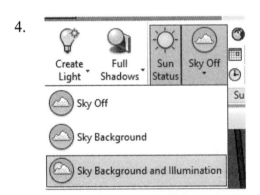 Select **Sky Background and Illumination** from the Render ribbon.

   If you haven't installed the hot fix, you will get an error dialog.

5.  In the drop-down panel under Sky Background and Illumination, select

   Set location→From Map.

6.
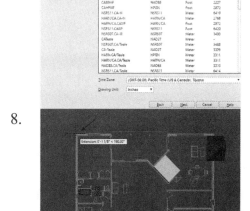

Type in an address in the search field where your project is located.

Once it is located on the map, select **Drop Marker here.**

Press **Next.**

7.

Select the coordinate system to be applied to the project.

Highlight **CA-III**, then press **Next.**

8.

Select the upper left corner of the building as the site location.

Press ENTER to accept the default North location.

9.

ACA has added a Geolocation ribbon with tools that allow you to modify the location, map the aerial position and create an image showing the placement of the building.

10.

The location of the project is overlaid on a satellite image.

11.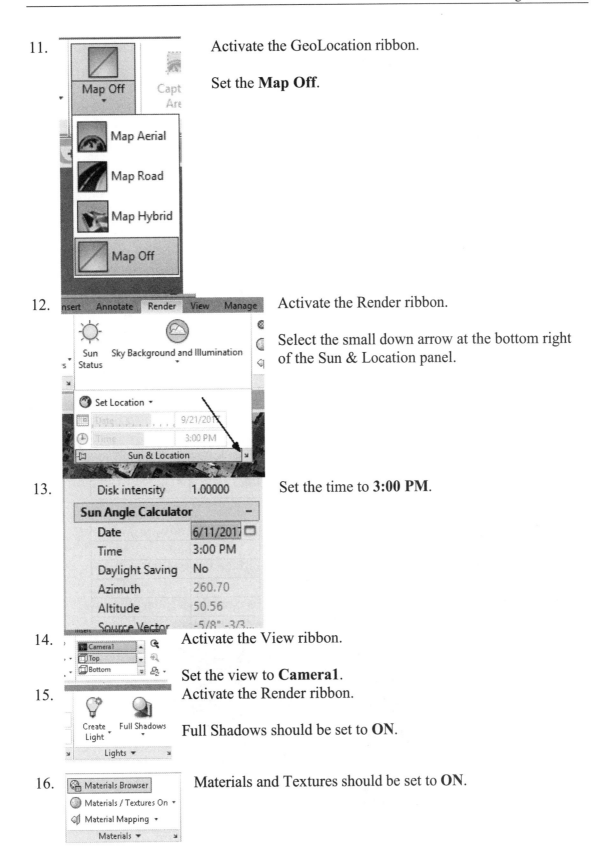

Activate the GeoLocation ribbon.

Set the **Map Off**.

12. Activate the Render ribbon.

Select the small down arrow at the bottom right of the Sun & Location panel.

13. Set the time to **3:00 PM**.

14. Activate the View ribbon.

Set the view to **Camera1**.

15. Activate the Render ribbon.

Full Shadows should be set to **ON**.

16. Materials and Textures should be set to **ON**.

17. On the Render to Size dropdown, select the **More Output Settings.**

18. Set the Image Size to 1280x720 px.
Set the Resolution to 600.
Press Save.

19.

Render to Size

Click on **Render to Size**.

If a dialog pops up regarding missing materials, just close it.

20. A window will come up where the camera view will be rendered.

If you don't see the window, use the drop-down to open the Render Window.

21. Select the **Save** icon to save the rendered image.

22. 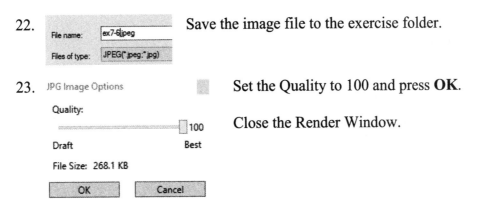   Save the image file to the exercise folder.

23. Set the Quality to 100 and press **OK**.

Close the Render Window.

24. Save as *ex7-6.dwg*.

Exercise 7-7:
Render in Cloud

Drawing Name: ex7-6.dwg
Estimated Time: 5 minutes

This exercise reinforces the following skills:

- ❑ Render in Cloud
- ❑ Autodesk 360

Autodesk offers a service where you can render your model on their server using an internet connection. You need to register for an account. Educators and students can create a free account. The advantage of using the Cloud for rendering is that you can continue to work while the rendering is being processed. You do need an internet connection for this to work.

1.   Open *ex7-6.dwg*.

2. 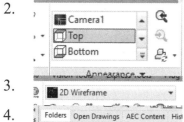 Activate the View ribbon.

   Switch to a **Top** View.

3.   Set the Display to **2D Wireframe**.

4.   Launch the Design Center by typing DC.

   Activate the Folders tab.

   Locate the D3D-P-LOW-001.dwg file in the exercise files.

   Highlight Blocks.

5.   Highlight the **D3D-P-LOW-001** block.

   Right click and select **Insert Block**.

6.  Set the Scale to **25.4**. Enable **Uniform Scale**.

   Enable **Specify On-Screen** Insertion Point.

   Press **OK**.

7.

Place the block inside the kitchen area.

This block is a point cloud, so it may be difficult to see.

8.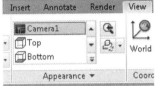

Switch to the Camera1 view.

9.

You will see the 3D figure inside the kitchen.

In order to render using Autodesk's Cloud Service, you need to sign in to an Autodesk 360 account. Registration is free and storage is free to students and educators.

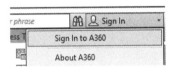

Sign in to your Autodesk 360 account.

10.

Enter your log-in information and press **Sign in**.

11. Your user name should display once you are signed in.

12. Save the file as *ex7-7.dwg*.

You must save the file before you can render.

13. Activate the Render ribbon.

Select **Render in Cloud**.

14.  Set the Model View to render the current view.

Click the **Start Rendering** model.

15. You should see a small window showing that the rendering operation is in process.

16. A small bubble message will appear when the rendering is completed.

17. You will receive an email when the rendering is complete.

18. Select **Render Gallery** on the Render ribbon to see your renderings.

19.

You can also select View completed rendering from the drop down menu under your A360 sign in.

Compare the quality of the rendering performed using the Cloud with the rendering done in the previous exercise.

I have found that rendering using A360 is faster and produces better quality images.

Exercise 7-8:
Create an Animation

Drawing Name: ex7-6.dwg
Estimated Time: 5 minutes

This exercise reinforces the following skills:

- ❑ Walk-Through
- ❑ Autodesk 360

1.  Open *ex7-6.dwg*.

2. 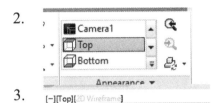 Activate the View ribbon.

   Switch to a **Top** View.

3. Set the view display to **2D Wireframe**.

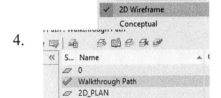

4. Activate the Home ribbon.

   Create a layer named **Walkthrough Path** and set it **Current**.

   This way you can freeze the path when it is not being used.

5. Select the **Polyline** tool located under the Line drop-down list.

6. Draw a polyline that travels through the floor plan to view different rooms.

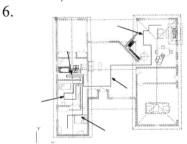

7.

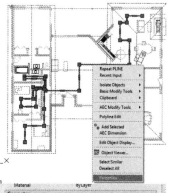

Select the polyline.

Right click and select **Properties**.

8.

Set the elevation of the polyline to **5' 3"**.

If you don't raise the line, your camera path is going to be along the floor.

9.

Activate the Render ribbon.

Animation Motion Path

Select **Animation Motion Path**.

10.

Set the Visual Style to **Realistic**.

Set the Format to **AVI**.
This will allow you to play the animation in Windows Media Player.

Set the Resolution to **1024 x 768**.

Set the Duration to **300 seconds**.

Enable **Path**.

Left click on the select button.
Select the polyline.

11.

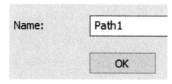

Name the camera path.

Press **OK**.

12.

File name: walkthrough1.avi

Files of type: AVI Animation (*.avi)

Press **OK**.

Browse to your classwork folder.

Name the file to create *walkthrough1.avi*.

Press **Save**.

13.

Frame number - 468 of 18000

22 Minutes, 0 Seconds Remaining

14.

You will see the animation preview as it processes.

This will take several minutes.
When it is complete, open Windows File Explorer.

Browse to the folder where it is saved.

Right click on it and select Open with Windows Media Player.

15. Try creating different walkthroughs using different visual styles.

16. Save as *ex7-8.dwg*.

Notes:

Lesson 8:
Documentation

Before we can create our construction drawings, we need to create layouts or views to present the design. AutoCAD Architecture is similar to AutoCAD in that the user can work in Model and Paper Space. Model Space is where we create the 3D model of our house. Paper Space is where we create or set up various views that can be used in our construction drawings. In each view, we can control what we see by turning off layers, zooming, panning, etc.

To understand paper space and model space, imagine a cardboard box. Inside the cardboard box is Model Space. This is where your 3D model is located. On each side of the cardboard box, tear out a small rectangular window. The windows are your viewports. You can look through the viewports to see your model. To reach inside the window so you can move your model around or modify it, you double-click inside the viewport. If your hand is not reaching through any of the windows and you are just looking from the outside, then you are in Paper Space or Layout mode.

You can create an elevation in your current drawing by first drawing an elevation line and mark, and then creating a 2D or 3D elevation based on that line. You can control the size and shape of the elevation that is generated. Unless you explode the elevation that you create, the elevation remains linked to the building model that you used to create it. Because of this link between the elevation and the building model, any changes to the building model can be made in the elevation as well.

When you create a 2D elevation, the elevation is created with hidden and overlapping lines removed. You can edit the 2D elevation that you created by changing its display properties. The 2D Section/Elevation style allows you to add your own display components to the display representation of the elevation, and create rules that assign different parts of the elevation to different display components. You can control the visibility, layer, color, linetype, lineweight, and linetype scale of each component. You can also use the line work editing commands to assign individual lines in your 2D elevation to display components and merge geometry into your 2D elevation.

After you create a 2D elevation, you can use the AutoCAD BHATCH and AutoCAD DIMLINEAR commands to hatch and dimension the 2D elevation.

AutoCAD Architecture comes with standard title block templates. They are located in the Templates subdirectory under *Program Data \ Autodesk\ ACA 2018 \enu \Template.*

Many companies will place their templates on the network so all users can access the same templates. In order to make this work properly, set the Options so that the templates folder is pointed to the correct path.

Exercise 8-1:

Creating a Custom Titleblock

Drawing Name: Architectural Title Block.dwg
Estimated Time: 30 minutes

This exercise reinforces the following skills:

- ❏ Title blocks
- ❏ Attributes
- ❏ Edit Block In-Place
- ❏ Insert Image
- ❏ Insert Hyperlink

1. 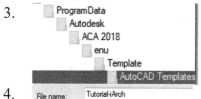   Select the **Open** tool.

2. File name: _____   Set the Files of type to *Drawing Template*.
    Files of type: Drawing Template (*.dwt)

3. ProgramData   Browse to
    Autodesk   *ProgramData\Autodesk\ACA 2018\enu*
    ACA 2018   *Template\AutoCD Templates*.
    enu
    Template
    AutoCAD Templates

4. File name: Tutorial-iArch   Open the *Tutorial-iArch.dwt*.
    Files of type: Drawing Template (*.dwt)   Verify that *Files of type* is set to Drawing Template or
    you won't see the file.

5. Perform a **File→Save as**.

    File name: Arch_D.dwt   Save the file to your work folder.
    Files of type: AutoCAD Drawing Template (*.dwt)   Rename **Arch_D.dwt**.

6.    Enter a description and press **OK**.

7.

Firm Name and Address

Zoom into the Firm Name and Address rectangle.

8.   **A**
     Multiline Text

Select the **MTEXT** tool from the **Annotation** panel on the Home ribbon.

Draw a rectangle to place the text.

9.   Set the Font to **Verdana**.

Hint: If you type the first letter of the font, the drop-down list will jump to fonts starting with that letter.

Set the text height to ⅛″.
Set the justification to **centered**.
Enter the name and address of your college.

10.

Extend the ruler to change the width of your MTEXT box to fill the rectangle.

Press **OK**.

11.   Firm Name and Address

LANEY COLLEGE
800 FALLON STREET
OAKLAND, CA

Use **MOVE** to locate the text properly, if needed.

12.  Activate the **Insert** ribbon.

Select the arrow located on the right bottom of the Reference panel.

13. The External Reference Manager will launch.

Select **Attach Image** from the drop-down list.

14. Locate the image file you wish to use.
There is an image file available for use from the publisher's website called *college logo.jpg*.
Press **Open**.

15. Press **OK**.

16.  Place the image in the rectangle.

If you are concerned about losing the link to the image file (for example, if you plan to email this file to another person), you can use **INSERTOBJ** to embed the image into the drawing. Do not enable link to create an embedded object. The INSERTOBJ command is not available on the standard ribbon.

---

17. Select the image.
    Right click and select **Properties**.

18.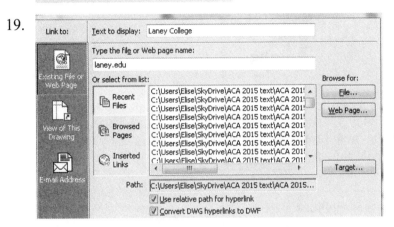

    Select the **Extended** tab.
    Pick the Hyperlink field.

19.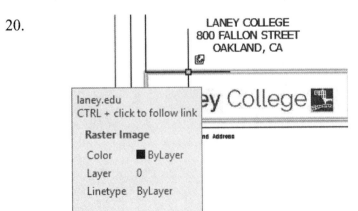

    Type in the name of the school in the Text to display field.

    Type in the website address in the *Type the file or Web page name* field.

    Press **OK**.
    Close the Properties dialog.

20.

    If you hover your mouse over the image, you will see a small icon. This indicates there is a hyperlink attached to the image. To launch the webpage, simply hold down the CTRL key and left click on the icon.

If you are unsure of the web address of the website you wish to link, use the Browse for **Web Page** button.

21.

Firm Name and Address

LANEY COLLEGE
800 FALLON STREET
OAKLAND, CA

To turn off the image frame/boundary,
type **IMAGEFRAME** at the command line.
Enter **0**.

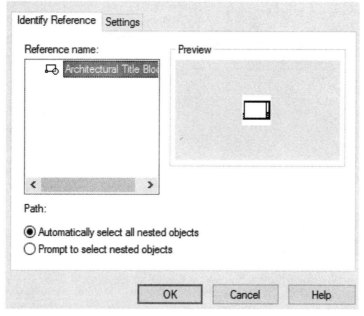

IMAGEFRAME has the following options:
  0: Turns off the image frame and does not plot
  1: Turns on the image frame and is plotted
  2: Turns on the image frame but does not plot

22.

Block Editor
Edit Block in-place...
Edit Attributes...

Select the titleblock.
Right click and select **Edit Block in-place**.

23.   **A** Reference Edit   ✕   Press **OK**.

Identify Reference | Settings

Reference name:

⊞ Architectural Title Blo

Preview

Path:

◉ Automatically select all nested objects
○ Prompt to select nested objects

[OK]   [Cancel]   [Help]

24.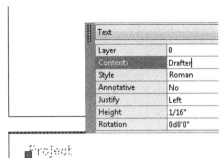

Change the text for Project to **Drafter**.
To change, simply double click on the text and an edit box will appear.

25. Select the **Define Attribute** tool from the Insert ribbon.

Define
Attributes

26. Select the **Field** tool.

27. Highlight Author and Uppercase.

Press **OK**.

28. In the Tag field, enter **DRAFTER**.
In the Prompt field, enter **DRAFTER**.
The Value field is used by the FIELD property.
   In the Insertion Point area:
      In the X field, enter **2′ 6.5″**.
      In the Y field, enter **2-1/16″**.
      In the Z field, enter **0″**.
   In the Text Options area:
      Set the Justification to **Left**.
      Set the Text Style to **Standard**.
      Set the Height to **1/8″**.

Press **OK**.

29. Select the **Define Attributes** tool.

Define
Attributes

30.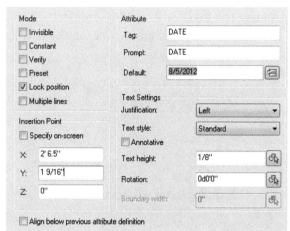

In the Tag field, enter **DATE**.
In the Prompt field, enter **DATE**.

In the Insertion Point area:
  In the X field, enter **2′ 6.5″**.
  In the Y field, enter **1-9/16″**.
  In the Z field, enter **0″**.
In the Text Options area:
  Set the Justification to **Left**.
  Set the Text Style to **Standard**.
  Set the Height to **1/8″**.

31.   Select the Field button to set the default value for the attribute.

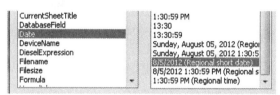

Select **Date**.
Set the Date format to **Regional short date**.

Press **OK**.

32.   Select the **Define Attributes** tool.

33.

In the Tag field, enter **SCALE**.
In the Prompt field, enter **SCALE**.
  In the Insertion Point area:
    In the X field, enter **2′ 6.5″**.
    In the Y field, enter **1-1/16″**.
    In the Z field, enter **0″**.
  In the Text Options area:
    Set the Justification to **Left**.
    Set the Text Style to **Standard**.
Set the Height to **1/8″**.

34.

Select the Field button to set the default value for the attribute.

Select **PlotScale**.
Set the format to **1:#**.

Press **OK**.

Press **OK** to place the attribute.

35.  Select the **Define Attributes** tool.

36.

In the Tag field, enter **SHEETNO**. In the Prompt field, enter **SHEET NO**.

   In the Insertion Point area:
      In the X field, enter **2′ 8.25″**.
      In the Y field, enter **1-9/16″**.
      In the Z field, enter **0**.
   In the Text Options area:
      Set the Justification to **Left**.
      Set the Text Style to **Standard**.
      Set the Height to **1/8″**.

37.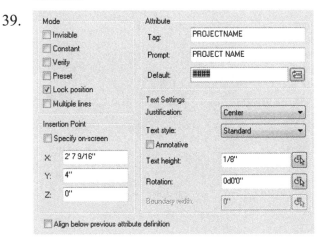

Select the Field button to set the default value for the attribute.

Select **CurrentSheetNumber**.
Set the format to **Uppercase**.

Press **OK**.

Press **OK** to place the attribute.

38. Select the **Define Attribute** tool.

39.

In the Tag field, enter **PROJECTNAME**.
In the Prompt field, enter **PROJECT NAME**.

   In the Insertion Point area:
      In the X field, enter **2′ 7-9/16″**.
      In the Y field, enter **4″**.
      In the Z field, enter **0**.
   In the Text Options area:
      Set the Justification to **Center**.
      Set the Text Style to **Standard**.
      Set the Height to **1/8″**.

40. Select the Field button to set the default value for the attribute.

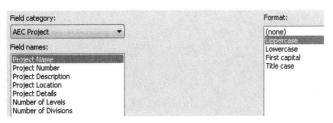

Select **AEC Project** for the field category.
Set the Field Name to **Project Name**.
Set the format to **Uppercase**.

Press **OK**.

A common error for students is to forget to enter the insertion point. The default insertion point is set to 0,0,0. If you don't see your attributes, look for them at the lower left corner of your title block and use the MOVE tool to position them appropriately.

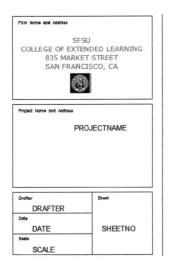

Your title block should look similar to the image shown.

If you like, you can use the MOVE tool to reposition any of the attributes to make them fit better.

41. 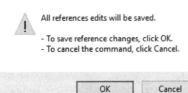 Select **Save** on the ribbon to save the changes to the title block and exit the Edit Block In-Place mode.

42. AutoCAD ☓ Press **OK**.

⚠ All references edits will be saved.

- To save reference changes, click OK.
- To cancel the command, click Cancel.

[OK] [Cancel]

43. Save the file and go to **File→Close**.

➤ Use templates to standardize how you want your drawing sheets to look. Store your templates on a server so everyone in your department uses the same titleblock and sheet settings. You can also set up dimension styles and layer standards in your templates.

➤ ADT 2005 introduced a new tool available when you are in Paper Space that allows you to quickly switch to the model space of a viewport without messing up your scale. Simply select the **Maximize Viewport** button located on your task bar to switch to model space.

➤ To switch back to paper space, select the **Minimize Viewport** button.

You use the Project Navigator to create additional drawing files with the desired views for your model. The views are created on the Views tab of the Project Navigator. You then create a sheet set which gathers together all the necessary drawing files that are pertinent to your project.

Previously, you would use external references, which would be external drawing files that would be linked to a master drawing. You would then create several layout sheets in your master drawing that would show the various views. Some users placed all their data in a single drawing and then used layers to organize the data.

This shift in the way of organizing your drawings will mean that you need to have a better understanding of how to manage all the drawings. It also means you can leverage the drawings so you can reuse the same drawing in more than one sheet set.

You can create five different types of views using the Project Navigator:

❏ Model Space View – a portion that is displayed in its own viewport. This view can have a distinct name, display configuration, description, layer snapshot and drawing scale.
❏ Detail View – displays a small section of the model, i.e., a wall section, plumbing, or foundation. This type of view is usually associated with a callout. It can be placed in your current active drawing or in a new drawing.
❏ Section View – displays a building section, usually an interior view. This type of view is usually associated with a callout. It can be placed in your current active drawing or in a new drawing.

- ❑ Elevation View – displays a building elevation, usually an exterior view. This type of view is usually associated with a callout. It can be placed in your current active drawing or in a new drawing.
- ❑ Sheet View – this type of view is created when a model space view is dropped onto a layout sheet.

Note: Some classes have difficulty using the Project Navigator because they do not use the same work station each class or the drawings are stored on a network server. In those cases, the links can be lost and the students get frustrated trying to get the correct results.

Exercise 8-2:

Verifying the Project Settings

Drawing Name: ex7-6.dwg
Estimated Time: 5 minutes

This exercise reinforces the following skills:

❑ Project Navigator
❑ Project Browser

1.   Open *ex7-6.dwg*.

2.   Right click in the command window and select **Options**.

3.   Select the **AEC Project Defaults** tab.

4.   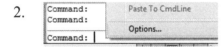 Verify that the AEC Project Location Search Path to the location where you are saving your files.

5.   Press **OK** to close the Options dialog.

6.   Launch the **Project Navigator** from the Quick Access toolbar.

7.

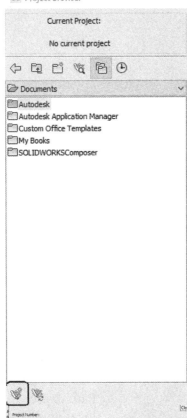

Select the **New Project** tool on the lower left of the dialog box.

8.

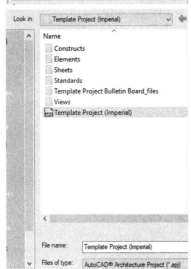

Type **P-101** for the Project Number.
Type **Brown Residence** for the Project Name.
Select the **Browse ...** button to select the template to use for the project.

9.

Locate the *Template Project (Imperial)* and select.

Press **OK** to close the Add Project dialog.

10. 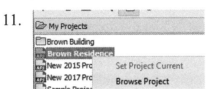 Verify that the Current Project is set to the Brown Residence project.

11. Right click on the **Brown Residence** project and select **Set Project Current**.

12. Save as *ex8-2.dwg*.

As we start building our layout sheets and views, we will be adding schedule tags to identify different building elements.

Schedule tags can be project-based or standard tags. Schedule tags are linked to a property in a property set. When you anchor or tag a building element, the value of the property displays in the tag. The value displayed depends on which property is defined by the attribute used by the schedule tag. ACA comes with several pre-defined schedule tags, but you will probably want to create your own custom tags to display the desired property data.

Property sets are specified using property set definitions. A property set definition is a documentation object that is tracked with an object or object style. Each property has a name, description, data type, data format, and default value.

To access the property sets, open the Style Manager. Highlight the building element you wish to define and then select the Property Sets button. You can add custom property sets in addition to using the default property sets included in each element style.

To determine what properties are available for your schedules, select the element to be included in the schedule, right click and select Properties.

Tags and schedules use project information. If you do not have a current project defined, you may see some error messages when you tag elements.

Exercise 8-3:

Creating Elevation Views

Drawing Name: ex8-2.dwg
Estimated Time: 15 minutes

This exercise reinforces the following skills:

- Creating an Elevation View
- Adding a Callout
- Named Views

1.  Open *ex8-2.dwg*.

2. Switch to the Model tab.

3. Activate the **Top** view by clicking the top plane on the view cube. *You can also use the short cut menu in the upper left corner of the display window to switch views.*

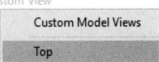

4.  Verify using the View cube that North is oriented on top.

5.  Set the view to **2D Wireframe** mode by selecting the View Display in the upper left corner of the window.

6.  Use **Zoom Extents** to view the entire model.

7.  Thaw the **A-Roof-Slab** and **A-Roof** layers.

8. Activate the **Annotate** ribbon.

   Select the **Elevation** tool.

9. Place the elevation mark below the model.
Use your cursor to orient the arrow toward the building model.

10.  Set your view name to **South Elevation**.

Enable **Generate Section/Elevation**.

Enable **Place Titlemark**.

Set the Scale to **1/8″ = 1′-0″**.

Press the **Current Drawing** button.

11. Window around the entire building to select it.
Select the upper left corner above the building and the lower right corner below the building.

12. 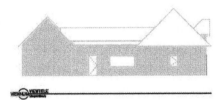 Place the elevation to the right of the view.

Zoom into the elevation view, so you can inspect it.

13. To activate the elevation view, select it from the View list located on the View ribbon.

14. Save as *ex8-3.dwg*.

Exercise 8-4:
Creating Layouts

Drawing Name: ex8-3.dwg
Estimated Time: 25 minutes

This exercise reinforces the following skills:

- ❑ Create Layout
- ❑ Adding a View to a Layout
- ❑ Control Visibility of elements using Layers

1.   Open *ex8-4.dwg*.

2.   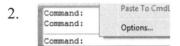 Right click on the command line and select **Options**.

You can also type OPTIONS.

3.   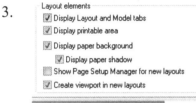 Activate the **Display** tab.

Enable **Display Layout and Model tabs**.

Press **OK**.

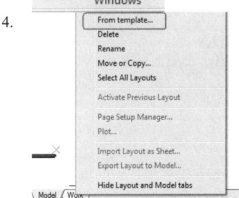

These can also be enabled on the View ribbon.

4.   Right click on the Work tab.

Select **From template**.

5.   Browse to your exercise folder.

Select the *Arch_D* template.
Press **Open**.

6.    Press **OK**.

7. `Model / Work \ D-Size Layout /`   Activate the **D-Size Layout**.

8.

| Rename |
| Move or Copy... |
| Select All Layouts |
| Activate Previous Layout |
| Activate Model Tab |
| Page Setup Manager... |
| Plot... |
| Import Layout as Sheet... |
| Export Layout to Model... |
| Hide Layout and Model tabs |

`Work \ D-Size Layout /`

Right click on the **D-Size Layout** tab.

Select **Rename**.

9. `\ Floorplan /`   Rename Floorplan.

10.

Viewport	
Layer	Viewport
Display locked	No
Annotation scale	1'-0" = 1'-0"
Standard scale	1/4" = 1'-0"
Custom scale	0"
Visual style	2D Wireframe
Shade plot	As Displayed

Select the viewport.
*If the Properties dialog doesn't come up, right click and select **Properties**.*

Set Display locked to **No**.

11.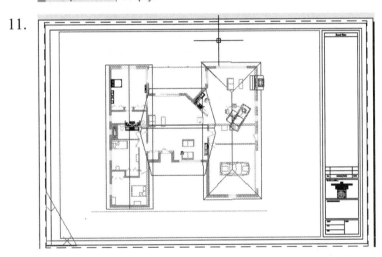

Resize the viewport to only show the floor plan.

Double click inside the viewport to activate model space.

Pan the model to position inside the viewport.

12. Activate the Home ribbon.

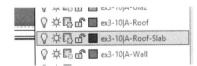

Freeze the layer assigned to roof slabs in the current viewport.

13.

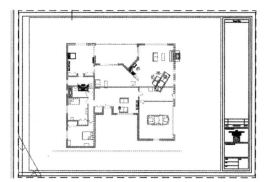

The view updates with no roof.

14. Activate the Home ribbon.

Freeze the layer assigned to the section in the current viewport.

To turn off the display of the camera, type **CAMERADISPLAY**, and then **0.**

15.

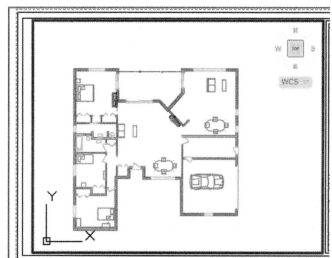

Double click outside the viewport to return to paper space.

Observe how the floor plan view has changed.

16.

Activate the Home ribbon.

Launch the Layer Manager.

17.

	S-Cols-Patt	Color_...		
	Title Block	Color_7		
►	Viewport	Color_...		

Scroll down to the Viewport layer.
Set the Viewport layer to DO NOT PLOT.

The viewports can be visible so you can adjust them but will not appear in prints.

Close the Layer Manager dialog.

18.

Misc	–
On	Yes
Clipped	No
Display locked	Yes
Annotation scale	1/4" = 1'-0"
Standard scale	1/4" = 1'-0"
Custom scale	1/32"
UCS per viewport	Yes
Layer property overrid...	No
Visual style	2D Wireframe
Shade plot	Hidden
Linked to Sheet View	No

Select the viewport.
On the Properties palette,
set the Annotation scale to **1/4″ = 1′-0″**,
set the standard scale to **1/4″ = 1′-0″**,
set the Shade plot to **Hidden**, and
set the Display locked to **Yes**.

19. Save as *ex8-4.dwg*.

Exercise 8-5:
Creating a Sheet

Drawing Name: ex8-4.dwg
Estimated Time: 15 minutes

This exercise reinforces the following skills:

- ❑ Layouts
- ❑ Viewports

1. Open *ex8-4.dwg*.

2. If your layout tabs are not displayed,
right click on the command line and select **Options**.

You can also type OPTIONS.

3. Layout elements
☑ Display Layout and Model tabs
☑ Display printable area
☑ Display paper background
   ☑ Display paper shadow
☐ Show Page Setup Manager for new layouts
☑ Create viewport in new layouts

Activate the **Display** tab.

Enable **Display Layout and Model tabs**.

Press **OK**.

4.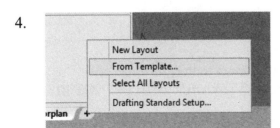

New Layout
From Template...
Select All Layouts
Drafting Standard Setup...

rplan

Right click on the + tab.

Select **From template**.

5. File name: Arch_D
Files of type: Drawing Template (*.dwt)

Browse to the folder where you are saving your work.
Select the *Arch_D* template.
Press **Open**.

6. 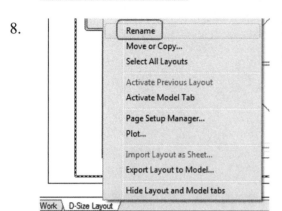 Press **OK**.

7. `Model / Work \ D-Size Layout /` Activate the **D-Size Layout**.

8.

Right click on the **D-Size Layout** tab.

Select **Rename**.

> Rename
> Move or Copy...
> Select All Layouts
>
> Activate Previous Layout
> Activate Model Tab
>
> Page Setup Manager...
> Plot...
>
> Import Layout as Sheet...
> Export Layout to Model...
>
> Hide Layout and Model tabs

`Work \ D-Size Layout /`

9. `Model / Work / Floorplan \ South Elevation /` Rename **South Elevation**.

10.

Layer	Viewport
Display locked	Yes
Annotation scale	Yes
Standard scale	No
Custom scale	0"

Select the viewport.
Right click and set the Display locked to **No**.

11. Double click inside the viewport to activate model space.

12.

Left click inside the viewport and then position the elevation view inside the viewport.

13.

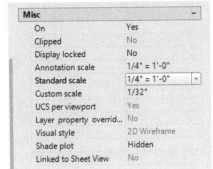

Select the viewport.
Set the Annotation scale to **1/4″ = 1′-0″ [1:100]**.
Set the Standard scale to **1/4″= 1′-0″ [1:100]**.

Set the Shade plot to **Hidden**.

You may need to re-center the view in the viewport after you change the scale. Use PAN to do this so the scale does not change.

Lock the display.

14. Save the file as *ex8-5.dwg*.

> By locking the Display you ensure your model view will not accidentally shift if you activate the viewport.

> The Annotation Plot Size value can be restricted by the Linear Precision setting on the Units tab. If the Annotation Plot Size value is more precise than the Linear Precision value, then the Annotation Plot Size value is not accepted.

Most projects require a floorplan indicating the fire ratings of the building walls. These are submitted to the planning department for approval to ensure the project is to building code. In the next exercises, we will complete the following tasks:

- Load the custom linetypes to be used for the exterior and interior walls
- Apply the 2-hr linetype to the exterior walls using the Fire Rating Line tool
- Apply the 1-hr linetype to the interior walls using the Fire Rating Line tool
- Add wall tags which display fire ratings for the walls
- Create a layer group filter to save the layer settings to be applied to a viewport
- Apply the layer group filter to a viewport
- Create the layout of the floorplan with the fire rating designations

Exercise 8-6:
Loading a Linetype

Drawing Name: ex7-3.dwg
Estimated Time: 10 minutes

This exercise reinforces the following skills:

    ❑   Loading Linetypes

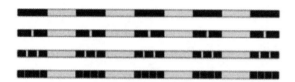

FIRE RATED, SMOKE BARRIER WALLS, 1 HOUR
FIRE RATED, SMOKE BARRIER WALLS, 2 HOUR
FIRE RATED, SMOKE BARRIER WALLS, 3 HOUR
FIRE RATED, SMOKE BARRIER WALLS, 4 HOUR

These are standard line patterns used to designate different fire ratings.

I have created the linetypes and saved them into the firerating.lin file included in the exercise files. You may recall we used the Express Tools Make Linetype tool to create custom linetypes.

1.   Open *ex7-3.dwg.*

2.   **Model**   Activate the Model tab.

3.   Type **LINETYPE** to load the linetypes.

4.   Select **Load**.

5.   Select the **File** button to load the linetype file.

6.   Locate the *fire rating.lin* file in the exercise files.

    Press **Open**.

7.   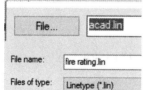   *This is a txt file you can open and edit with Notepad.* Hold down the control key to select both linetypes available.

    Press **OK** to load.

8.   The linetypes are loaded.

    Press **OK** to close the dialog box.

9.   Save as *ex8-6.dwg.*

Exercise 8-7:
Applying a Fire Rating Line

Drawing Name: ex8-6.dwg
Estimated Time: 25 minutes

This exercise reinforces the following skills:

- Adding a Layout
- Adding Fire Rating Lines

1. Open *ex8-6.dwg*.

2.
   New Layout
   From Template...
   Select All Layouts
   Drafting Standard Setup...
   Model | Work | +

   Right click on the + layout tab.

   Select **From Template**.

3.
   File name: Arch_D.dwt

   Files of type: Drawing Template (*.dwt)

   Select the **Arch_D** template from the exercise folder.

   Press **Open**.

4.
   Layout name(s):   OK

   D-Size Layout   Cancel

   Highlight the Layout name.

   Press **OK**.

5.
   Delete
   Rename
   Move or Copy...
   Select All Layouts
   Activate Previous Layout
   Activate Model Tab
   Page Setup Manager...
   Plot...
   Drafting Standard Setup...
   Import Layout as Sheet...
   Export Layout to Model...
   Dock above Status Bar
   D-Size L...

   Right click on the D-Size Layout.

   Select **Rename**.

   Rename to **Floorplan with Fire Rating**.

6.

   Launch the Layer Manager.

   Set the Viewport layer to **Do Not Plot**.

7. Launch the **Design Tools** palette from the Home ribbon if it is not available.

8. Right click on the Design Tools palette title bar.

Enable **Document**.

9. Double left click inside the view port to activate model space.

10. 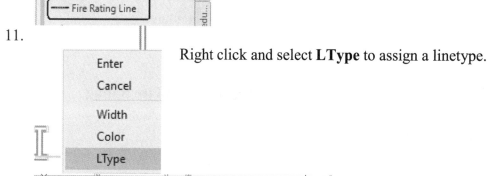 Select the Annotation tab.

Locate and select the **Fire Rating Line** tool.

11. Right click and select **LType** to assign a linetype.

12. `ANNORATINGLINEADD Specify polyline linetype or [?]<Aec_Rating_4Hr>: 2-hr`

Type **2-hr**.
Press **Enter**.

13. Right click and select **Width**.

14. 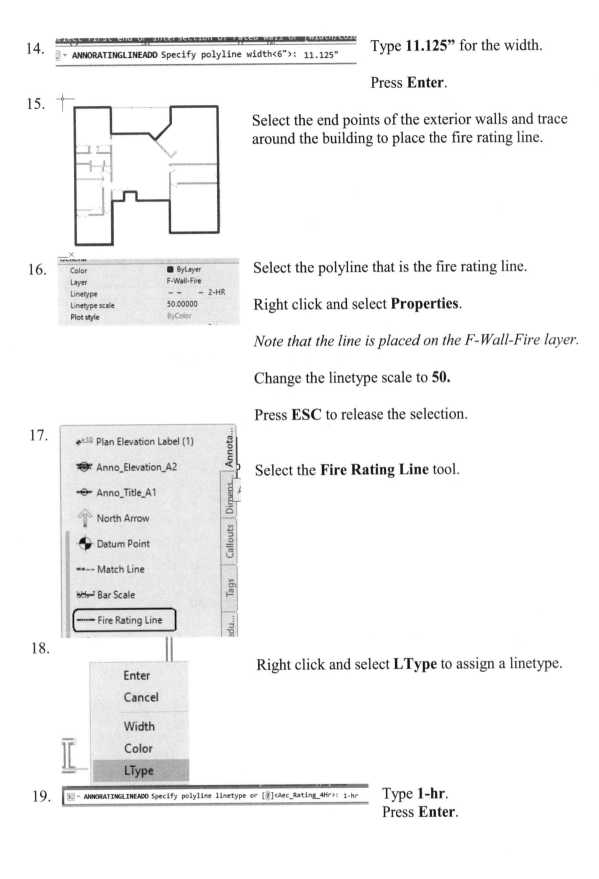 ▾ **ANNORATINGLINEADD** Specify polyline width<6">: 11.125"

Type **11.125"** for the width.

Press **Enter**.

15. Select the end points of the exterior walls and trace around the building to place the fire rating line.

16. Select the polyline that is the fire rating line.

Right click and select **Properties**.

Note that the line is placed on the F-Wall-Fire layer.

Change the linetype scale to **50**.

Press **ESC** to release the selection.

17. Select the **Fire Rating Line** tool.

18. Right click and select **LType** to assign a linetype.

19. ▾ **ANNORATINGLINEADD** Specify polyline linetype or [?]<Aec_Rating_4Hr>: 1-hr

Type **1-hr**.
Press **Enter**.

20.

        Enter

        Cancel

        Recent Input

        Width

        Color

        LType

Right click and select **Width**.

21. 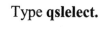 `ANNORATINGLINEADD Specify polyline width<6">: 6.5"`

Type **6.5"**.
Press **Enter**.

22.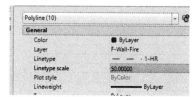

Place the 1-hr polyline on all the interior walls.

You will need to place more than one polyline as the interior walls are not all connected.

23. Type **qslelect.**

Select **Polyline** from the Object Type drop-down list.

Highlight **Linetype** in Properties.

Set the Value to **1-hr.**

Press **OK.**

24. Right click and select **Properties**.

Set the Linetype scale to **50**.

25. Click outside the viewport to return to paper space.

Save as *ex8-7.dwg*.

Exercise 8-8:
Assigning Fire Ratings to Wall Properties

Drawing Name: ex8-7.dwg
Estimated Time: 10 minutes

This exercise reinforces the following skills:

❏ Wall Properties

1. Open *ex8-7.dwg*.

2. Activate the *Floor Plan with Fire Rating* layout.

    rk   **Floor Plan with Fire Rating**   +

    Double click inside the viewport to activate model space.

3. Activate the Manage ribbon.

    Style Manager

    Select the **Style Manager**.

4.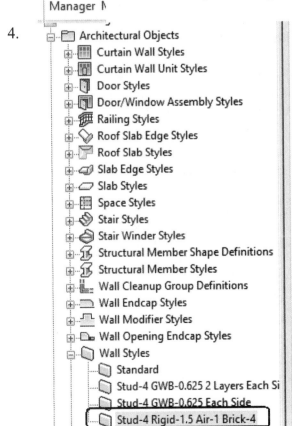

    Locate and select the **Stud-4 Rigid- 1.5 Air-1-Brick-4** Wall Style.

    This is the wall style used for the exterior walls.

5.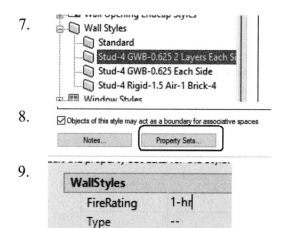

    On the General tab:

    Select **Property Sets**.

    Property sets control the values of parameters which are used in schedules and tags.

6.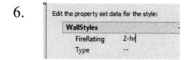

    Type **2-hr** in the FireRating property.

    Press **OK**.

    When defining the attribute to use this value, we use the format WallStyles:FireRating. The first word is the property set and the second word is the property.

7. [Wall Styles tree showing: Wall Opening Endcap Styles, Wall Styles, Standard, Stud-4 GWB-0.625 2 Layers Each S[, Stud-4 GWB-0.625 Each Side, Stud-4 Rigid-1.5 Air-1 Brick-4, Window Styles]

    Highlight the **Stud-4 GWB-0.625 2 Layers Each Side** Wall Style.

    This is the wall style used for the interior walls.

8. ☑ Objects of this style may act as a boundary for associative spaces

    [Notes...] [Property Sets...]

    On the General tab:

    Select **Property Sets**.

    Type **2-hr** in the FireRating property.

    Press **OK**.

9. **WallStyles**

    FireRating   1-hr

    Type   --

10. Close the Styles Manager.

    Double click outside the viewport to return to paper space.

    Save the file as *ex8-8.dwg*.

Exercise 8-9:
Creating a Schedule Tag

Drawing Name: ex8-8.dwg
Estimated Time: 30 minutes

This exercise reinforces the following skills:

- ❏ Wall Tags
- ❏ Create a Schedule Tag
- ❏ Create a Tool on the Tool Palette
- ❏ Property Sets

1. Open *ex8-8.dwg*.

2. 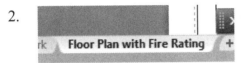 Activate the Floor Plan with Fire Rating layout.

    Double click inside the viewport to activate model space.

3.  Freeze the **F-Wall-Fire** layer.

    This will make it easier to select walls.

4. 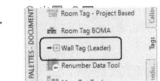 Activate the Tags tab on the Documentation palette.

    Select the **Wall Tag** tool.

5. 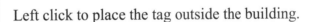 Select an exterior wall.

    Left click to place the tag outside the building.

    Browse through the property sets and note that the FireRating is set to 2-HR.

    Press **OK**.

6. 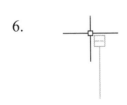 If you zoom into the end of the leader line for the wall tag, you will see the tag. It is very small and doesn't have the correct attribute for the fire rating.

    We will create a new block definition, so we can see the wall tag and it shows the fire rating.

    Select this wall tag and delete it.

7. 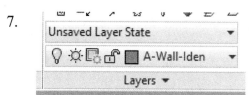 Set the **A-Wall-Iden** layer current.

This is the layer to be used by the new wall tag.

8. Draw a 10' 8" x 10' 8" square.

9.  Select the square.
Right click and select **Basic Modify Tools→Rotate**.

When prompted for the base point, right click and select **Geometric Center**.

Then select the center of the square.

Type **45** for the angle of rotation.

10. Activate the Insert ribbon.

Select the **Define Attributes** tool.

Define Attributes

11. Type **WALLSTYLES:FIRERATING** for the tag.
This is the property set definition used for this tag.
Type **Fire Rating** for the prompt.
Type **2-hr** for the Default value.

Set the Justification for the **Middle center**.
Enable **Annotative**.
Set the Text Height to **2'**.

Enable Specify on-screen.

Press **OK**.

12.  Select the Geometric Center of the Square to insert the attribute.

13.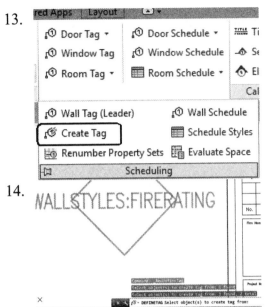

Activate the Annotate ribbon.

Select the **Create Tag** tool from the Scheduling panel in the drop-down area.

14.

Select the square and the attribute you just created.

Press **ENTER**.

15.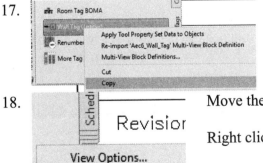

Name the new tag **FireRating_Wall_Tag**.

You should see a preview of the tag and you should see the Property Set and Property Definition used by the tag based on the attribute you created.

Press **OK**.

16.

The block will be created and the tag will re-size based on the view scale.

Select the block and delete it.

Save the drawing as *ex8-9.dwg*.

17.

Locate the Wall Tag (Leader) tool on the tool palette.

Right click and select **Copy**.

18.

Move the cursor to a blank space on the palette.

Right click and select **Paste**.

19. 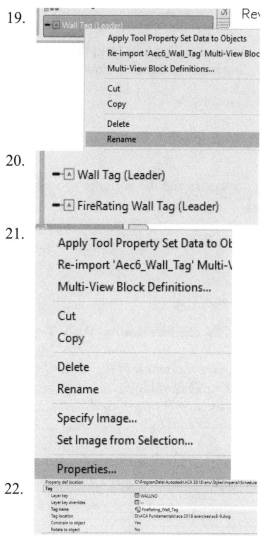 Highlight the copied Wall Tag tool.

Right click and select **Rename**.

20. Rename it **FireRating Wall Tag (Leader)** and move it below the original wall tag.

21. Highlight the **FireRating Wall Tag.**

Right click and select **Properties.**

22. Set the Tag Location as *ex8-9.dwg.*
Set the Tag Name to **FireRating_Wall_Tag.**

You may want to create a drawing to store all your custom schedule tags and then use that drawing when setting the tag location.

23.

Press **OK**.
You may need to re-zoom to see your floor plan. The viewport is locked so it will not affect the layout.

Select the **FireRating Wall Tag** tool and tag an exterior wall.

24.

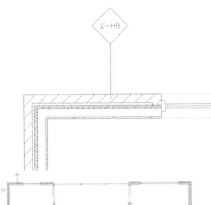

Left click to place the tag and then press **OK** to accept the properties.

You should see the updated tag.

25.

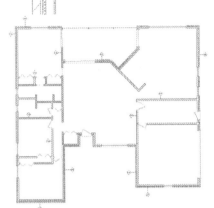

Go through the floor plan to tag all the walls.

Double click outside the viewport to switch to paper space.

Save as *ex8-9.dwg*.

Exercise 8-10:

Modify a Dimension Style

Drawing Name: ex8-9.dwg
Estimated Time: 15 minutes

This exercise reinforces the following skills:

- ❏ Dimension Styles
- ❏ AEC Dimensions

1. Open *ex8-9.dwg*.

2. 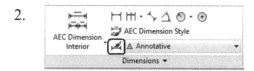 Activate the Annotate ribbon.

   Select the **Dimension Style Editor** on the Dimensions panel.

3. Highlight **Annotative**.

4. Select **Modify**.

5. On the Lines tab,

   set the Baseline spacing to **1/2″**.

6. Set Extend beyond dim lines to **1/8″**.

   Set Offset from origin to **1/4″**.

7.  On the Symbols and Arrows tab,

   set the Arrow size to **1/4″**.

8.   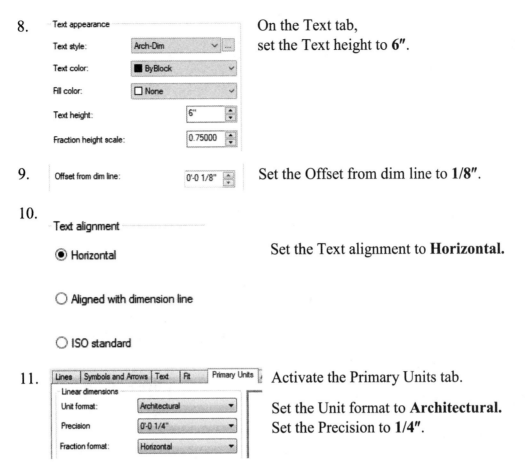 On the Text tab,
set the Text height to **6″**.

9.   Set the Offset from dim line to **1/8″**.

10.   Set the Text alignment to **Horizontal.**

11.   Activate the Primary Units tab.

Set the Unit format to **Architectural.**
Set the Precision to **1/4″**.

12.   Press **OK** and **Close**.

13.   Save as *ex8-10.dwg*.

Exercise 8-11:

Dimensioning a Floor Plan

Drawing Name: ex8-10.dwg
Estimated Time: 30 minutes

This exercise reinforces the following skills:

- ❑   Drawing Setup
- ❑   AEC Dimensions
- ❑   Title Mark
- ❑   Attributes
- ❑   Layouts

1.   Open *ex8-10.dwg*.

2.   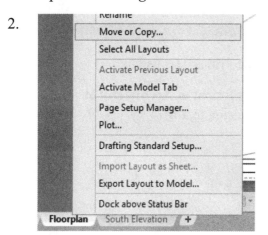   Highlight the **Floor Plan with FireRating** layout tab.

Right click and select **Move or Copy**.

3.   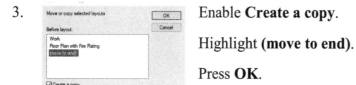   Enable **Create a copy**.

Highlight **(move to end)**.

Press **OK**.

4. | Model / Work / Floorplan / South Elevation / Floorplan (2) /   Select the second layout.

5.   Highlight the second layout tab.

Right click and select **Rename**.

6. 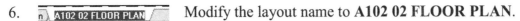 Modify the layout name to **A102 02 FLOOR PLAN**.

7.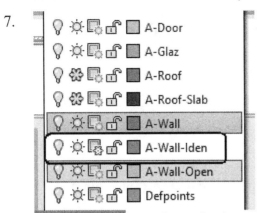

Activate model space for the new layout.

Activate the Home ribbon.

Locate the **A-Wall-Iden** layer on the layer drop-down list.

Select **Freeze in Viewport** to turn off the wall tags.

8. 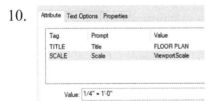 Activate the Annotate ribbon.

Select the **Title Mark** tool.
Pick below the view to place the title mark.

9.

Pick two points to indicate the start and end points of the title mark.

Place the title mark below the view.

10.

Tag	Prompt	Value
TITLE	Title	FLOOR PLAN
SCALE	Scale	ViewportScale

Value: 1/4" = 1'-0"

Double click to edit the attributes.
In the Title field, enter **FLOOR PLAN**.
In the SCALE field, enter ¼" = 1'-0".

Press **OK**.

11. Text Style: Annotative
Justification: Left   Backwards   Upside down
Height: 1'-0"   Width Factor: 1"
Rotation: 0.00   Oblique Angle: 0.00
Annotative

Highlight the Title tag.
Select the Text Options tab.
Set the Text Height to 1'.
Press Apply to see a preview.
Repeat for the Scale tag.
Press **OK**.

12.

The view title updates.

Use the grips for the tags to re-position the text above and below the line.

13. 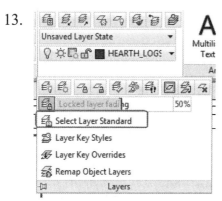   A   Activate the Home ribbon.

In the Layers panel drop-down,
select **Select Layer Standard**.

14.    Set the Layer Key Style to **AIA (256 Color)**.

Press **OK**.

15. 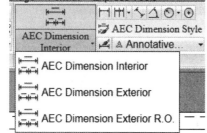   Select the **AEC Dimension - Exterior** tool on the
Annotate ribbon.

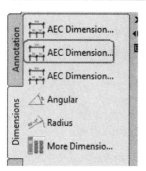   This tool is also available on the Dimensions tab of the
Annotation tool palette.

*Add Wall Dimensions is only available in Model Space. Dimensions are
automatically placed on A-Anno-Dim layer when the Layering Standard is set to
AIA.*

16.

Select the walls, doors, and windows indicated on the right side of the building.

Press ENTER to complete the selection.

17.

Pick to the right of the building to place the dimensions.

All dimensions relevant to the wall will be placed.

18. Select the Dimension you just placed.

Remove Ext. Lines

Select **Remove Extension Lines** from the ribbon.

19.

Select the grips on the extension lines that are not necessary.
You should see a tool tip indicating the extension line to remove.

Remove extension Line

The selected lines will be deleted.

20. To add an extension line,
select the AEC Dimension.

Select the Add Extension Lines tool from the ribbon. Add Extension Lines
Then select the midpoint of the object to be dimensioned.
Then select the AEC Dimension.

21.   When you are done modifying the AEC
    Dimension, press ENTER to update the
    dimensions.

22.   Select the dimension.

*The dimension should automatically be placed on A-Anno-Dims because the Layer
Standard was set to AIA.*

23. Switch back to Paper space.

    PAPER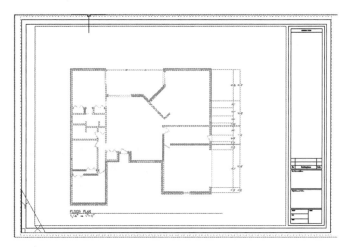

24. Save as *ex8-11.dwg*.

Tips
Tricks

You can insert blocks into model or paper space. Note that you did not need to
activate the viewport in order to place the Scale block.

Exercise 8-12:

Create a Wall Section

Drawing Name: ex8-11.dwg
Estimated Time: 15 minutes

This exercise reinforces the following skills:

- ❑ Wall Section
- ❑ Layouts

Most building documents include section cuts for each wall style used in the building model. The section cut shows the walls from the roof down through the foundation. Imagine if you could cut through a house like a cake and take a look at a cross section of the layers, that's what a wall section is. It shows the structure of the wall with specific indication of materials and measurements. It doesn't show every wall, it is just a general guide for the builder and the county planning office, so they understand how the structure is designed.

1. Open *ex8-11.dwg*.

2. 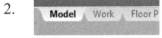 Activate the Model layout tab.

3.  Freeze the **A-Wall-Iden** layer.

4. Activate the **Annotate** ribbon.

    Select the **Section (Sheet Tail)** tool.

5. Draw the section through the south wall of the garage area.

    Select a point below the wall.
    Select a point above the wall.
    Right click and select ENTER.
    Left click to the right of the section line to indicate the section depth.

6.

Type Exterior Wall Section in the View Name field.

Enable Current Drawing.

Enable Generate Section/Elevation.

Enable Place Title mark.
Set the Scale to 1/8″ = 1′-0″.

7.

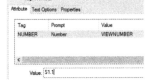

Place the view to the right of the South Elevation view.

8.

Select the view title for the wall section.

Type **1/8″ = 1′-0″** for the SCALE.

Type **Exterior Wall Section** for the TITLE.

Press ESC to release and update.

9.

Select the View title bubble.

Type **S1.1** for the NUMBER.
Press **OK**.

Press **ESC** to release and update.

10.

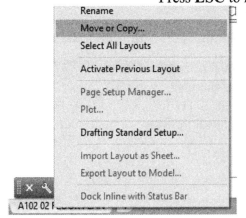

Highlight the **A102 02 Floor Plan** layout tab.

Right click and select **Move or Copy**.

11. 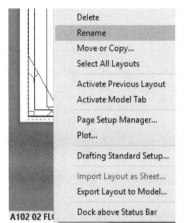 Enable **Create a copy**.

Highlight **(move to end)**.

Press **OK**.

11. Select the second layout.

12. Highlight the second layout tab.

Right click and select **Rename**.

13. Modify the layout name to **S1.2 Details**.

14. Unlock the viewport.
Position the Exterior Wall Section in the viewport.
Adjust the size of the viewport.

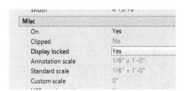 Select the viewport.
Set the Annotation scale to **1/8″ = 1′0″**.
Set the Standard scale to **1/8″ = 1′0″**.

Set the Shade plot to **As Displayed**.

Lock the display.

The title bar will not display unless the view is set to the correct scale.

15. Press ESC to release the viewport.

16. Save as *ex8-12.dwg*.

Exercise 8-13:

Add Keynotes

Drawing Name: ex8-12.dwg
Estimated Time: 15 minutes

This exercise reinforces the following skills:

- ❑ Keynote Database
- ❑ Adding Keynotes

Keynotes are used to standardize how materials are called out for architectural drawings. Most keynotes use a database. The database is created from the Construction Specifications Institute (CSI) Masterformat Numbers and Titles standard. ACA comes with several databases which can be used for keynoting or you can use your own custom database. The database ACA uses is in Microsoft Access format (*.mdb). If you wish to modify ACA's database, you will need Microsoft Access to edit the files.

1. Open *ex8-12.dwg.*

2.  Activate the **S1.2 Details** layout tab.

    Double click inside the wall section viewport to activate model space.

3.  *Verify that the keynote database is set properly in Options.*

    Type OPTIONS to launch the Options dialog.

    Select the AEC Content tab.

    Select the Add/Remove button next to **Keynote Databases**.

4.  Verify that the paths and file names are correct.

    Press **OK**.

5.  Activate the Annotate ribbon.

    Select the **Reference Keynote (Straight)** tool on the Keynoting panel.

6.  Select the outside component of the wall section.

    The Keynote database dialog window will open.

7.

Locate the keynote for the **Clay/facing brickwork**.

Highlight and press **OK**.

8.

Select the outside component to start the leader line.
Left click to locate the text placement.
Press ENTER to complete the command.

9.

Repeat to place the keynote for the air barrier.

10.

To turn off the shading on the keynotes, turn off the field display.

Type **FIELDDISPLAY** and then **0**.

11.

Locate the keynote for the rigid insulation and add to the view.

Note that if you select the component on the wall section, ACA will automatically take you to the correct area of the database to locate the proper keynote.

12.

Locate the keynote for the 2x4 Studs and add to the view.

13.

Locate the keynote for the 5/8" Gypsum Board and add to the view.

14.

Save as *ex8-13.dwg*.

Exercise 8-14:
Create a Section View

Drawing Name: ex7-8.dwg
Estimated Time: 10 minutes

This exercise reinforces the following skills:

❏   Sections

Section views show a view of a structure or building as though it has been sliced in half or along an imaginary plane. This is useful because it gives a view through spaces and can reveal the relationships between different parts of the building which may not be apparent on the plan views. Plan views are also a type of section view, but they cut through the building on a horizontal plane instead of a vertical plane.

1.   Open *ex7-8.dwg*.

2.   Activate the **Model** layout tab.
        Activate Model space.

3.   Freeze the **A-Roof** and **A-Roof-Slab** layers.

4.   Turn off the visibility of the geomarker.

        Type **GEOMARKERVISIBILITY**.
        Type **0**, then **ENTER**.

5.   Activate the Annotate ribbon.

        Select the **Section (Sheet Tail)** tool on the Callouts panel.

6.   Draw a horizontal line through the master bathroom walls as shown.

7.

Name the View **Master Bath Elevation**.
Enable Generate Section/Elevation.
Enable Place Titlemark.
Set the Scale to **1/2″ = 1′-0″**.

Place the view in the current drawing to the right of the building.

8.

Right click on the + tab.

Select **From Template**.

9.

Select the **Arch_D** template from the classwork folder.

Press **Open**.

10.

Select the **D-size Layout**.

Press **OK**.

11.

Highlight the layout tab.

Right click and select **Rename**.

12.

Rename the layout name to Master Bath Elevation.

13.

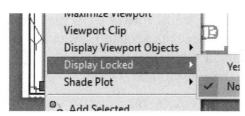

Unlock the viewport display.
Select the Viewport, right click and set Display Locked to No.

14.

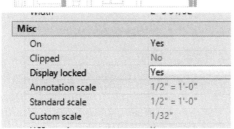

Activate Model space by double clicking inside the viewport.

Position and resize the viewport to show the Master Bath Elevation.

15.

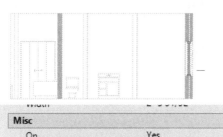

Select the viewport and set the scales to ½" = 1'-0".

Hint: You will not see the title bar unless the annotation scale matches the viewport scale and is set correctly. Remember this is the scale you selected when you created the view.

Set the Display Locked to **Yes**.

16.

Change the values of the view title attributes as shown.

17. Save as *ex8-14.dwg*.

To create an elevation view with different visual styles, create a camera view.

Callouts are typically used to reference drawing details. For example, builders will often reference a door or window call-out which provides information on how to frame and install the door or window. Most architectural firms have a detail library they use to create these call-outs as they are the same across building projects.

Exercise 8-15:

Add a Callout

Drawing Name: ex8-13.dwg
Estimated Time: 45 minutes

This exercise reinforces the following skills:

- ❑ Callouts
- ❑ Viewports
- ❑ Detail Elements
- ❑ Layouts
- ❑ Hatching
- ❑ Block Editor

1. Open *ex8-13.dwg*.

2.  Activate the **Model** tab.

3. 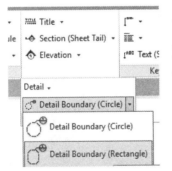 Zoom into the laundry room area.

   We will create a callout for the interior door frame.

4. 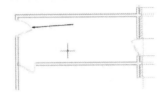 Activate the Annotate ribbon.

   Locate and select the **Detail Boundary (Rectangle)** tool on the Detail drop-down list on the Callouts panel.

5. 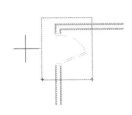 Draw a small rectangle to designate the callout location around the door.

6. 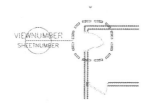 Select the midpoint of the left side of the rectangle for the start point leader line.

Left click to the side of the callout to place the label.

Press **ENTER** or right click to accept.

7. 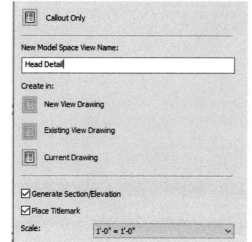 Type **Standard Interior Door Frame** in the View Name field.

Enable **Generate Section/Elevation**.

Enable **Place Titlemark**.

Set the Scale to **1'-0″ = 1'-0″**.

Click on the **Current Drawing** button.

8. 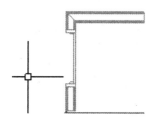 Press ENTER to accept the default elevation for the view.

9. Press ENTER to accept the default depth for the view.

10. Place the view to the right of the section view.

11. Activate the **S1.2 Details** layout tab.

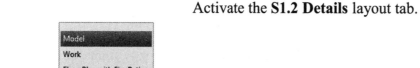

12.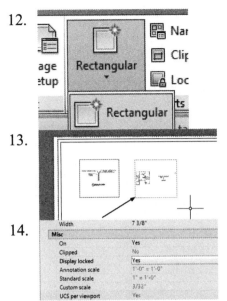

Activate the Layout ribbon.

Select the **Rectangular** viewport tool.

13.

Draw a rectangle by selecting two opposite corners to place the viewport.

14.

Set the Viewport scale to 1'-0" = 1'-0".

Adjust the size of the viewport.

Position the door view in the viewport.

Lock the display.

15.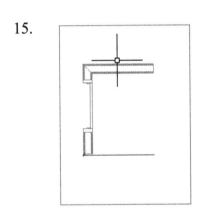

Place the viewport on the viewport layer.

16.

Double click inside the viewport to activate model space.

Select the view title for the door callout.

Type **1'-0" = 1'-0"** for the SCALE.

Type **Head Detail – Interior Door** for the TITLE.

17.

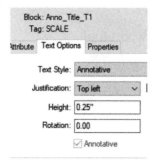

Select the Text Options tab.

Change the Text Height to 0.25" for both attributes.

Press **OK**.

Press **ESC** to release and update.

18.

Block: Anno_Title_A1
Tag: NUMBER

Attribute Text Options Properties

Tag	Prompt	Value
NUMBER	Number	D12

Value: D12

Select the View title bubble.

Type **D12** for the NUMBER.

Select the Text Options tab.

Change the Text Height to **0.25"**.

Press **ESC** to release and update.

19.

Block Editor
Edit Block in-place...
Edit Attributes...

Select the bubble.

Right click and select **Block Editor**.

20.

Geometry
Center X	0"
Center Y	0"
Center Z	0"
Radius	4"
Diameter	8"

Select the circle.

Right click and select **Properties.**

Change the Diameter of the circle to **8"**.

21.

Save
Block

Press **Save Block** on the ribbon.

22.

Close
Block Editor

Close

Select **Close Block Editor** on the ribbon.

23.

The changes you made have not been saved. What do you want to do?

→ Save the changes to Anno_Title_A1

→ Discard the changes and close the Block Editor

Select Save the changes.

24. 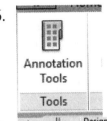 Use the grips to position the text and the title line so it looks correct.

25. 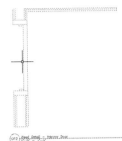 Adjust the sizes of the text and block until you are happy with the title bar.

26. 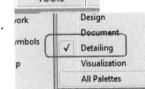 Launch the tool palette.

Annotation Tools

Tools

27. Right click on the Tool palette's bar.

Enable the Detailing palette.

28. 06 - Woods, Plastics & Composites   Select the **06- Woods, Plastics % Composites** detail tool on the Basic tab.

29. Right click and select Rotate to rotate the detail.

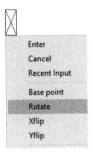

Enter
Cancel
Recent Input
Base point
Rotate
Xflip
Yflip

30.  Place the stud in the upper door frame location.

31. Select the stud detail.

Right click and select Properties.
Change the X scale to **1.3**.
Change the Y scale to **1.87**.

The stud size will adjust to fit the frame.

To get those values, I took the desired dimension 6.5 and divided it by the current dimension 3.5 to get 1.87. I took the desired dimension of 2.0 and divided it by the current dimension of 1.5 to get 1.3. To determine the dimensions, simply measure the current length and width and the desired length and width.

32. Copy and place the stud to the other side of the door.

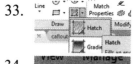

33. Activate the Home ribbon.

Select the **Hatch** tool.
34. Select the **ANSI31** hatch.

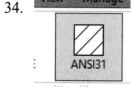

35. Place a hatch in the three areas indicated.

Close the Hatch Creation ribbon.

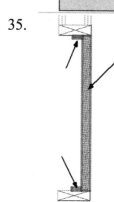

36.

Text (Straight Leader) ▾

Use the Text (Straight Leader) tool to add the text as shown.

Text should read:

ADD RETURN ON FRAME WHERE DRYWALL FINISH IS REQUIRED

37.

Use Properties to adjust the size of the arrowhead and the text.

Set the Arrowhead size to ½".
Set the Text Height to 1".

38.

Select the text.
Right click and select **Add Leader**.

39.

Add a second leader to the other side of the door frame.

40.

Select the Linear Dimension tool on the Annotate ribbon.

Place a dimension and modify the text to read:

THROAT OPENING = PARTITION WIDTH + 1/8" MIN.

41. Double click outside of the viewport to return to paper space.
42. Activate the A102 02 Floor Plan layout tab.

A102 02 FLOOR PLAN

Double click inside the viewport to activate model space.

43. Update the attributes in the callout with the correct sheet number S1.2 and Detail Number D12.

D12
S1.2

If you do not see the view titles on the layout, type ANNOALLVISIBLE and set to 1.

44. Save as *ex8-15.dwg*.

Space Styles

Depending on the scope of the drawing, you may want to create different space styles to represent different types of spaces, such as different room types in an office building. Many projects require a minimum square footage for specific spaces – such as the size of bedrooms, closets, or common areas – based on building code. Additionally, by placing spaces in a building, you can look at the traffic flow from one space to another and determine if the building has a good flow pattern.

You can use styles for controlling the following aspects of spaces:

- Boundary offsets: You can specify the distance that a space's net, usable, and gross boundaries will be offset from its base boundary. Each boundary has its own display components that you can set according to your needs.

- Name lists: You can select a list of allowed names for spaces of a particular style. This helps you to maintain consistent naming schemes across a building project.

- Target dimensions: You can define a target area, length, and width for spaces inserted with a specific style. This is helpful when you have upper and lower space limits for a type of room that you want to insert.

- Displaying different space types: You can draw construction spaces, demolition spaces, and traffic spaces with different display properties. For example, you might draw all construction areas in green and hatched, and the traffic areas in blue with a solid fill.

- Displaying different decomposition methods: You can specify how spaces are decomposed (trapezoid or triangular). If you are not working with space decomposition extensively, you will probably set it up in the drawing default.
Space Styles are created, modified, or deleted inside the Style Manager.

Exercise 8-16:
Creating Space Styles

Drawing Name: Ex8-15.dwg
Estimated Time: 10 minutes

1. Open *ex8-15.dwg*.

2.  Activate the **Manage** ribbon.

   Select the **Style Manager**.

3. Locate the *Space Styles* inside the drawing.

   Make sure you set the Files of Type to dwg or you won't see it.

   Note that there is only one space style currently available in the *ex8-15.dwg*.

4. **Style Manager**

   File Edit View

   Select the **Open Drawing** tool.

5.  Browse to the exercise folders for this text.
   Change the files of type to **Drawing (*.dwg)**.
   Open the *space styles.dwg*.

   File name: space styles
   Files of type: Drawing (*.dwg)

6. Expand the *space styles.dwg* in the Style Manager.

   Note several space styles have been defined.

   - Space Styles
     - Bathroom
     - Bedroom
     - Closet
     - Den
     - Dining Room
     - Garage
     - Hallway
     - Kitchen
     - Livingroom
     - Standard

7.  Highlight the **Space Styles** in the *space styles.dwg*.
    Right click and select **Copy**.

8.  Scroll up to the *ex8-15.dwg*.

    Highlight **Space Styles**.

    Right click and select **Paste**.

9. 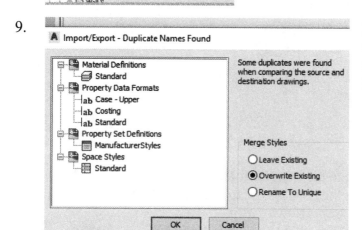 If this dialog appears advising of duplicate definitions,
    enable **Overwrite Existing**.
    Press **OK**.

10. 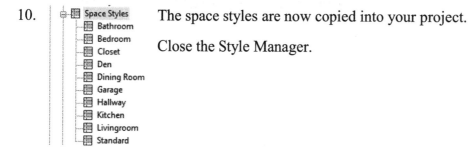 The space styles are now copied into your project.

    Close the Style Manager.

11. Save as *ex8-16.dwg*.

Exercise 8-17:

Adding Spaces

Drawing Name: Ex8-16.dwg
Estimated Time: 15 minutes

This exercise reinforces the following skills:

- ❑ Layouts
- ❑ Spaces

1.   Open *ex8-16.dwg*.

2.   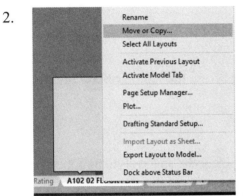

Right click on the A102 02 Floorplan layout tab and select **Move or Copy**.

Enable **Create a copy**.

Highlight **(move to end)**.

Press **OK**.

3.      Select the second layout.

4.   Highlight the second layout tab.

Right click and select **Rename**.

5.   Modify the layout name to **A102 03 ROOM PLAN**.

6. 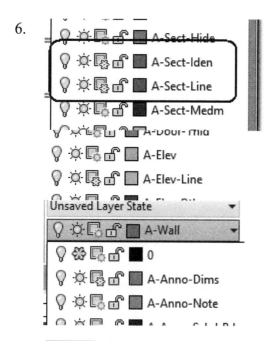 Double click inside the viewport to activate model space.

Freeze in Viewport the following layers:

A-Sect-Iden
A-Sect-Line
A-Elev-Line
A-Anno-Dims

7. 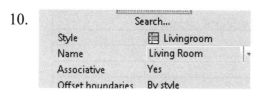 Select the **Generate Space** tool from the Home ribbon on the Build panel.

8.  On the Properties palette, select the space style to apply.

Left click inside the designated space to place the space.

9. If you hover over a space, you should see a tool tip with the space style name.

10. Select a space.

In the Properties palette select the Name for the SPACE using the drop down list.

This is the name that will appear on the room tag.

11. Repeat for all the spaces.
Press ESC to release the selection of each space before moving on to the next space.

Verify that only one space is selected to be renamed each time or you may assign a name to the wrong space unintentionally.

12. Save as *ex8-17.dwg.*

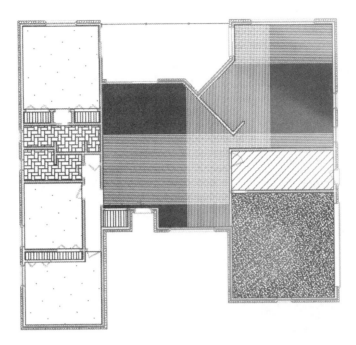

Exercise 8-18:
Modifying Spaces

Drawing Name: Ex8-17.dwg
Estimated Time: 15 minutes

This exercise reinforces the following skills:

❑   Spaces

1.   Open *ex8-17.dwg.*

2.

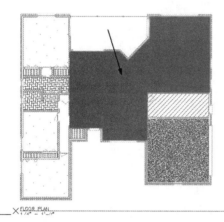

Select the space located in the living room area.

3.

On the Space contextual ribbon, select **Divide Space**.

4.

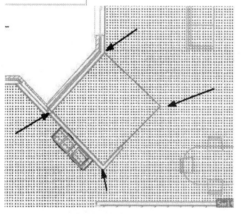

Turn off Object Snap to make selecting the points easier.

Place points to define a kitchen area.

5.

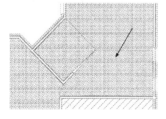

Repeat the Divide Space command to place lines to define the den and dining room areas.

Make sure each divided space is a closed polygon.

6.

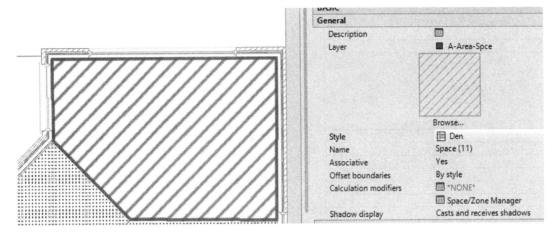

Select the den area space.

Right click and select **Properties**.

Change the Style to **Den.**
Change the Name to **Den.**

Press **ESC** to release the selection.

7.

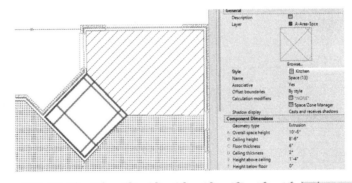

Select the kitchen area space.

Right click and select **Properties**.

Change the Style to **Kitchen.**
Change the Name to **Kitchen**.

Press **ESC** to release the selection.

8.

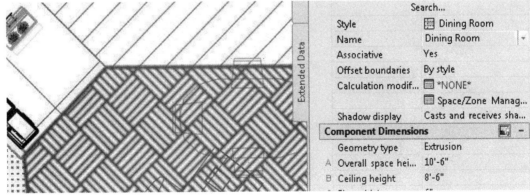

Select the dining room area space.

Right click and select **Properties**.

Change the Style to **Dining Room.**
Change the Name to **Dining Room.**

Press **ESC** to release the selection.

9.

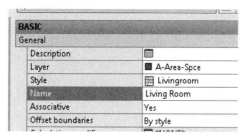

Select the Living room space.
Right click and select Properties.
On the Design tab,
change the name to **Living Room**.

Repeat for the remaining spaces.

13.

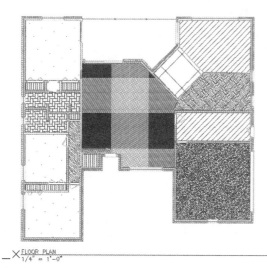

Inspect the floor plan to verify that all spaces have been placed.

Note that different space styles use different hatch patterns.

14. Save as *ex8-18.dwg*.

Exercise 8-19:

Adding Room Tags

Drawing Name: ex8-18.dwg
Estimated Time: 20 minutes

This exercise reinforces the following skills:

- ❑ Tags
- ❑ Tool Palette
- ❑ Content Browser

1.   Open *ex8-18.dwg*.

2.   Launch the Annotation Tools palette.

3.   Activate the Tags tab on the tool palette.

       Select **More Tag Tools**.

4.    Select the **Global Catalog**.

5.   Select **Documentation**.

6.   Select **Schedule Tags**.

7.   Locate the **Space Tag**.

       Drag and drop it onto the Tags tool palette.

8.   Select the **A102 03 ROOM PLAN** layout tab.
       Double click in the viewport to activate model space.

9.  Activate the **A102 03 Room Plan** layout.
Double click inside the view port to activate model space.

10. 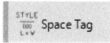 Select the **Space Tag** on the tool palette.

11.  Select the family room space.
Press **ENTER** to place the tag centered in the room.

12.  A small dialog will appear showing all the data attached to the space.

You can modify any field that is not grayed out.

Press **OK**.

13.
Den

244.74 SF
19'–7" x 13'–9"

Zoom in to see the tag.

Right click and select **Multiple** and then window around the entire building.

This will add space tags to every space.

You will be asked if you want to re-tag spaces which have already been tagged.
Select No.

14.

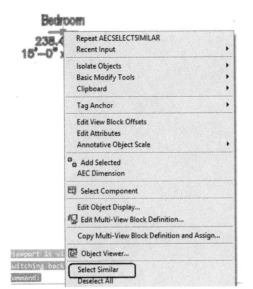

Zoom into one of the space tags.

Select it.

Right click and select **Select Similar**.

15.

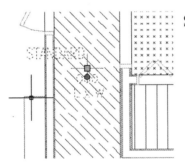

All the space tags will be selected.
Right click and select **Properties**.

Set the X scale to **25**.

Press ESC to release the selection.

16.

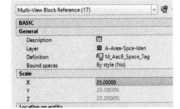

Select the Hallway tag.

17.

Location on entity

Anchor

On the Properties palette,
select **Anchor**.

18.

Rotation

Angle: 90

Set the Rotation angle to **90**.
Press **OK**.

19. Press ESC or left click outside the building model to release the selected tag.

20. 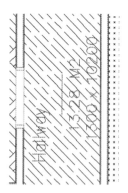 The tag is rotated 90 degrees.

21.   Save as *ex8-19.dwg*.

Exercise 8-20:

Adding Door Tags

Drawing Name: ex8-19.dwg
Estimated Time: 15 minutes

This exercise reinforces the following skills:

❑   Tags

1.   Open *ex8-19.dwg*.

2.   `n ) A102 02 Floor Plan ( S` Activate the **A102 02 Floor plan** layout tab.

3.   Activate model space by double-clicking inside the viewport.

4.   ☿ ☼ ▣ ⌂ ■ A-Area-Spce   Activate the Home ribbon.
       ☿ ☼ ▣ ⌂ ■ A-Area-Spce-Iden   Freeze the following layers in the current viewport.
       ☿ ☼ ▣ ⌂ ■ A-Area-Spce-Patt

                              A-Area-Spce
                              A-Area-Spce-Iden
                              A-Area-Spce-Patt

5.   Activate the Annotate ribbon.
       Select the Door Tag tool.

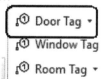

6.   Select a door.

       Press ENTER to center the tag on the door.

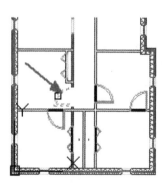

7.

Edit the property set data for the object:

PROPERTY SETS	−
DoorObjects	−
Data source	C:\Users\Elise\SkyDr...
Number	01
ƒₓ NumberProject...	"Space not found"A
RoomNumber	"Space not found"
NumberSuffix	A
Stule	Hinged - Single

Press **OK**.

Some students may be concerned because they see the Error getting value next to the Room Number field.

If you tag the spaces with a room tag assigning a number, you will not see that error.

8. Right click and select **Multiple**. Window around the entire building.

All the doors will be selected.

Press **ENTER**.

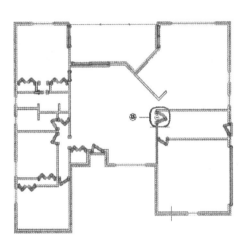

9.

1 object was already tagged with the same tag. Do you want to tag it again?

Press **No**.

| Yes | No |

10. Edit the property set data for the objects:

Press **OK.**

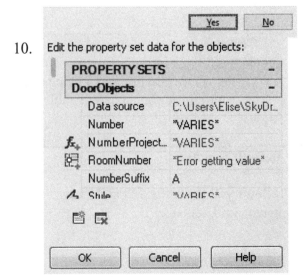

PROPERTY SETS	−
DoorObjects	−
Data source	C:\Users\Elise\SkyDr...
Number	"VARIES"
ƒₓ NumberProject...	"VARIES"
RoomNumber	"Error getting value"
NumberSuffix	A
Stule	"VARIES"

| OK | Cancel | Help |

11.

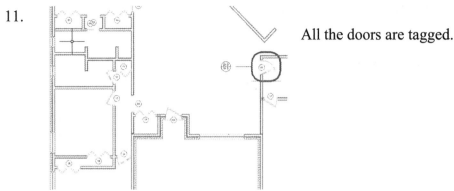

All the doors are tagged.

12. Save as *ex8-20.dwg*.

Schedule Styles

A schedule table style specifies the properties that can be included in a table for a specific element type. The style controls the table formatting, such as text styles, columns, and headers. To use a schedule table style, it must be defined inside the building drawing file. If you copy a schedule table style from one drawing to another using the Style Manager, any data format and property set definitions are also copied.

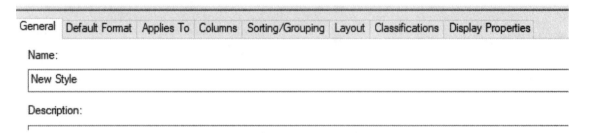

When you create a new Schedule Table Style, there are eight tabs that control the table definition.

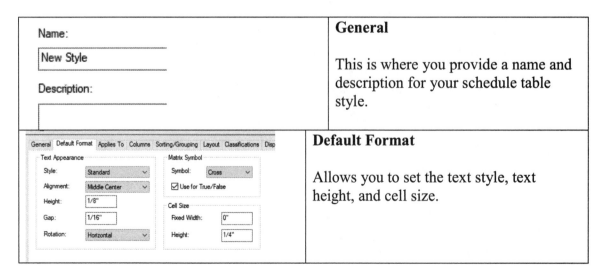

	General
	This is where you provide a name and description for your schedule table style.
	Default Format
	Allows you to set the text style, text height, and cell size.

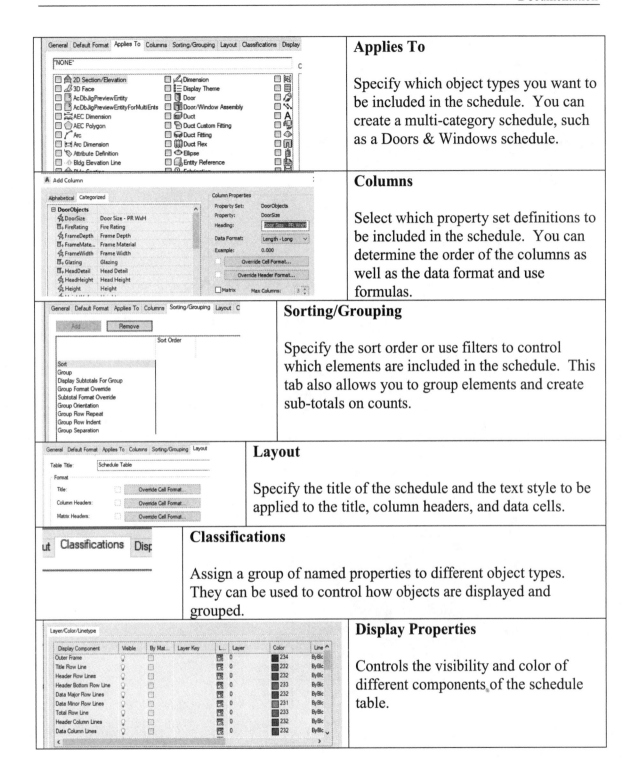

Applies To

Specify which object types you want to be included in the schedule. You can create a multi-category schedule, such as a Doors & Windows schedule.

Columns

Select which property set definitions to be included in the schedule. You can determine the order of the columns as well as the data format and use formulas.

Sorting/Grouping

Specify the sort order or use filters to control which elements are included in the schedule. This tab also allows you to group elements and create sub-totals on counts.

Layout

Specify the title of the schedule and the text style to be applied to the title, column headers, and data cells.

Classifications

Assign a group of named properties to different object types. They can be used to control how objects are displayed and grouped.

Display Properties

Controls the visibility and color of different components of the schedule table.

Exercise 8-21:

Create a Door Schedule

Drawing Name: ex8-20.dwg
Estimated Time: 10 minutes

This exercise reinforces the following skills:

❑   Schedules

1.   Open *ex8-20.dwg*.

2.   **A102 02 FLOOR PLAN**   Activate the **A102 02 FLOOR PLAN** layout tab.

Double click inside the viewport to activate model space.

3.   Door Schedule   Activate the Annotate ribbon.

Select the **Door Schedule** tool.

4.   Window around the building model.

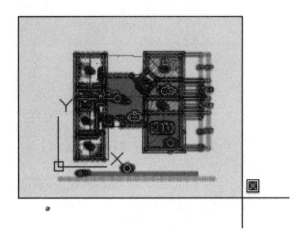

ACA will automatically filter out any non-door entities and collect only doors.

Press **ENTER**.

5.

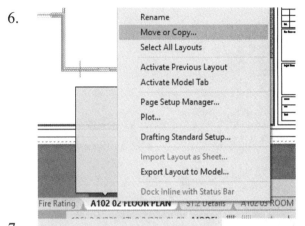

Left click to place the schedule in the drawing window.

Press **ENTER**.

6.

Rename
Move or Copy...
Select All Layouts

Activate Previous Layout
Activate Model Tab

Page Setup Manager...
Plot...

Drafting Standard Setup...

Import Layout as Sheet...
Export Layout to Model...

Dock Inline with Status Bar

Fire Rating   A102 02 FLOOR PLAN   S1.2 Details   A102 03 ROOM

Highlight the Floor Plan.

Right click and select **Move or Copy**.

7.

Move or copy selected layouts

Before layout:

Work
Floor Plan with Fire Rating
A102 02 FLOOR PLAN
S1.2 Details
A102 03 ROOM PLAN
(move to end)

☑ Create a copy

Enable **Create a copy**.

Highlight **(move to end)**.

Press **OK**.

8.

De
Re
Mc
Sel

Ac
Ac

Pa
Plc

Dr

Im
Ex

Do

AN   A102 02 FLOOR PLAN (2)

Highlight the copied layout.

Right click and select **Rename**.

9.

AN   S101 01 DOOR SCHEDUL

Rename **S101 01 DOOR SCHEDULE**.

10.   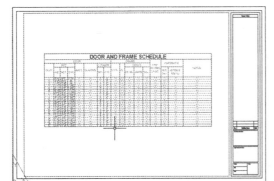   Unlock the display and position the door schedule centered in the viewport.

You may need to move the door schedule to an area of the drawing with more space.

11.   *A ? will appear for any doors that are missing tags. So, if you see a ? in the list go back to the model and see which doors are missing tags.*

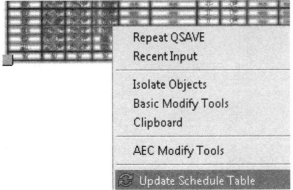

Once you add door tags, you need to update the schedule. Select the schedule, right click and select Update Schedule Table.

12.   `tion ⟍ A102 02 Floor Plan ⟋ S1.2 Details ⟋`   Activate the **A102 02 Floor Plan** layout tab.

13.   ⬚ Rectangular ▾   Activate the Layout ribbon.
Select the **Rectangular** viewport tool.

14. 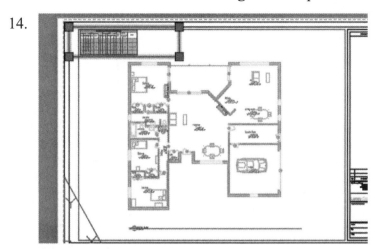   Draw a rectangular viewport.

Position the door schedule in the viewport.

15.

Viewport	▾
Layer	Viewport
Display locked	Yes
Annotation scale	1/8" = 1'-0"
Standard scale	Custom
Custom scale	0"
Visual style	2D Wireframe
Shade plot	Hidden

Select the viewport.

Verify that it is on the Viewport layer.

Set the Standard scale to **Custom.**

Set the Display locked to **Yes**.

16. Save as *ex8-21.dwg*.

Exercise 8-22:

Create a Door Schedule Style

Drawing Name: ex8-21.dwg
Estimated Time: 30 minutes

This exercise reinforces the following skills:

- ❑ Schedules
- ❑ Style Manager
- ❑ Text Styles
- ❑ Update Schedule

1.   Open *ex8-21.dwg*.

2.   `tion ⟩ A102 02 Floor Plan ⟨ S1.2 Details ⟩` Activate the **A102 02 Floor Plan** layout tab.

3.   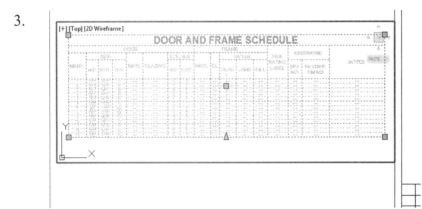   Double click inside the viewport with the door schedule.

Select the schedule.

4.   Right click and select **Copy Schedule Table Style**.

This will create a new style that you can modify.

5.   Select the **General** tab.

Rename **Custom Door Schedule**.

6.   `es To │ Columns │ Sorting` Select the **Columns** tab.

7. 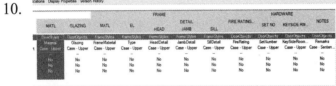 Highlight Louver, WD, and HGT.

8. [Delete...] Press **Delete**.

9. **A** Remove Columns/Headers

You will be asked to confirm the deletion.

Press **OK**.

☑ HGT(DoorStyles:LouverHeight)
☑ LOUVER
☑ WD(DoorStyles:LouverWidth)

[OK]   [Cancel]

10. Delete the columns under frame and hardware.

11. **A** Remove Columns/Headers   ✕

Press **OK**.

☑ FRAME
☑ GLAZING(DoorObjects:Glazing)
☑ HARDWARE
☑ HEAD(FrameStyles:HeadDetail)
☑ JAMB(FrameStyles:JambDetail)
☑ KEYSIDE RM NO(DoorObjects:KeySideRoom)
☑ MATL(DoorStyles:Material)
☑ MATL(FrameStyles:FrameMaterial)
☑ NOTES(DoorObjects:Remarks)
☑ SET NO(DoorObjects:SetNumber)
☑ SILL(FrameStyles:SillDetail)

[OK]   [Cancel]

12. 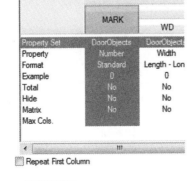 Highlight **MARK**.

Select the **Add Column** button.

13.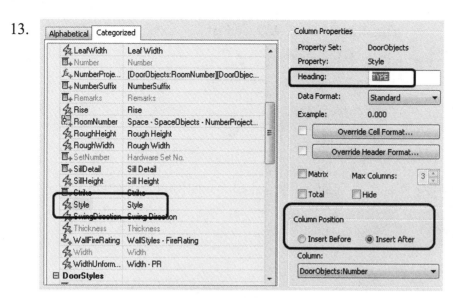

Scroll down until you see Style.

Highlight **Style**.

Confirm that the Column Position will be inserted after the DoorObjects: Number. Change the Heading to **TYPE**.

Press **OK**.

14.

The new column is now inserted.

15.

Select the Layout tab.

Change the Table Title to **DOOR SCHEDULE**.

16.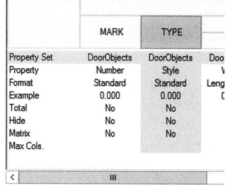

Activate the Columns tab.

Highlight the **WD** column.

Select **Modify**.

17.

Column Properties	
Property Set:	DoorObjects
Property:	Width
Heading:	WIDTH
Data Format:	Length - Long
Example:	0.000

Change the Heading to **WIDTH**.

Press **OK**.

18.

SIZE	
HGT	THK
DoorObjects	DoorObject
Height	Thickness
Length - Long	Length - Shc
0.000	0.0000
No	No
No	No
No	No

Quantity Column

Modify...

Highlight the **HGT** column.

Select **Modify**.

19.

Column Properties	
Property Set:	DoorObjects
Property:	Height
Heading:	HEIGHT
Data Format:	Length - Long
Example:	0.000

Change the Heading to **HEIGHT**.

Press **OK**.

20. Delete the THK column.

21. Remove Columns/Headers

☑ THK(DoorObjects.Thickness)

OK   Cancel

Press **OK**.

22. Press **OK** to close the dialog.

23.

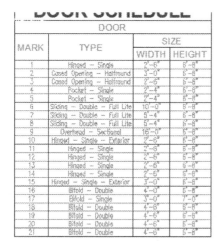

The schedule updates.

If there are question marks in the schedule, that indicates doors which remain untagged.

24.

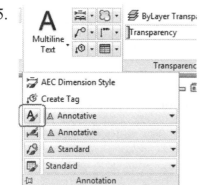

Select the schedule.
Select **Update** from the ribbon.
Press **ESC** to release the selection.

25.

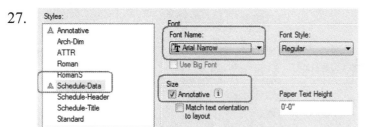

Activate the Home ribbon.

Select the **Text Style** tool on the Annotation panel.

26.

RomanS
Schedule-Data
Schedule-Header
Schedule-Title

Notice there are text styles set up for schedules.
Highlight each one to see their settings.

27.

Highlight **Schedule-Data**.
Set the Font Name to **Arial Narrow**.
Enable **Annotative**.

Press **Apply**.

28.

If this dialog appears, press **Yes**.

29. 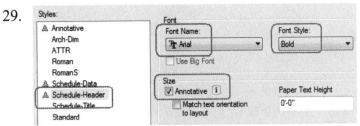 Highlight Schedule-Header. Set the Font Name to **Arial**. Set the Font Style to **Bold**. Enable **Annotative**.

Press **Apply**.

30. 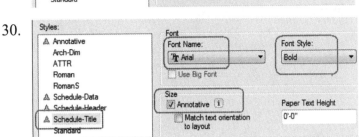 Highlight Schedule-Title. Set the Font Name to **Arial**. Set the Font Style to **Bold**. Enable **Annotative**.

Press **Apply** and **Close**.

31. Note that the schedule automatically updates.

DOOR SCHEDULE			
	DOOR		
		SIZE	
MARK	TYPE	WIDTH	HEIGHT
1	Hinged - Single	2'-8"	6'-8"
2	Cased Opening - Halfround	3'-0"	6'-8"
3	Cased Opening - Halfround	2'-6"	6'-8"
4	Pocket - Single	2'-4"	6'-8"
5	Pocket - Single	2'-4"	6'-8"
6	Sliding - Double - Full Lite	10'-0"	6'-8"
7	Sliding - Double - Full Lite	6'-4"	6'-8"
8	Sliding - Double - Full Lite	6'-4"	6'-8"
9	Overhead - Sectional	10'-0"	6'-8"
10	Hinged - Single - Exterior	2'-8"	6'-8"
11	Hinged - Single	2'-8"	6'-8"
12	Hinged - Single	2'-8"	6'-8"
13	Hinged - Single	2'-6"	6'-8"
14	Hinged - Single	2'-0"	6'-8"
15	Hinged - Single - Exterior	3'-0"	6'-8"
16	Bifold - Double	4'-0"	6'-8"
17	Bifold - Single	3'-0"	7'-0"
18	Bifold - Double	4'-0"	6'-8"
19	Bifold - Double	4'-0"	6'-8"
20	Bifold - Double	4'-6"	6'-8"
21	Bifold - Double	4'-6"	6'-8"

32. Save as *ex8-22.dwg*.

Exercise 8-23:
Create a Room Schedule

Drawing Name: ex8-22.dwg
Estimated Time: 30 minutes

This exercise reinforces the following skills:

- Schedules
- Tool Palettes
- Creating Tools
- Schedule Styles

1. Open *ex8-22.dwg*.

2. Activate the **A102 03 ROOM PLAN** layout tab.

   Double click inside the viewport to activate model space.

3. Select **Room Schedule** from the Annotation ribbon.

4. Window around the building model.
   Press ENTER.
   Pick to place the schedule.
   Press ENTER.

5. Activate the Manage ribbon.

   Select **Style Manager**.

6. Highlight the Schedule Table Styles category.

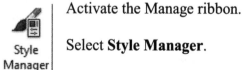

7. Right click and select **New**.
   On the General tab:

   Type **Room Schedule** for the Name.

   Type **Uses Space Information** for the Description.

8.

Select the Applies To tab.

Enable **Space**.

Note that Spaces have Classifications available which can be used for sorting.

9.

Select the Columns tab.

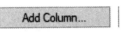

Select **Add Column**.

10.

Select **Style**.

Change the Heading to **Room Name**.

Change the Data Format to **Case – Sentence**.

Press **OK**.

11.

Select **Add Column**.

12.

Select the Alphabetical tab to make it easier to locate the property definition.

Locate and select **Length**.

Press **OK**.

13.

Select **Add Column**.

14.

Select the Alphabetical tab to make it easier to locate the property definition.

Locate and select **Width**.

Press **OK**.

15.

Select Add Column.

16.

Select the Alphabetical tab to make it easier to locate the property definition.

Select **BaseArea**.

Press **OK**.

17. 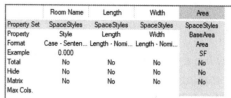 The columns should appear as shown.

	Room Name	Length	Width	Area
Property Set	SpaceStyles	SpaceStyles	SpaceStyles	SpaceStyles
Property	Style	Length	Width	BaseArea
Format	Case - Senten...	Length - Nomi...	Length - Nomi...	Area
Example	0.000			SF
Total	No	No	No	No
Hide	No	No	No	No
Matrix	No	No	No	No
Max Cols.				

18. Select the **Sorting/Grouping** tab.
Highlight **Sort**.
Select the **Add** button.
Select **SpaceStyles: Style** to sort by Style name.
Press **OK**.

19. Place a check next to **Group**.
Place a check next to **Display Subtotals for Group.**

Press **OK** to close the dialog.

20. 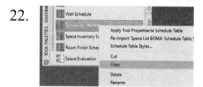 Launch the Annotation Tools palette from the Annotate ribbon.

21. Select the Schedules tab.

22. 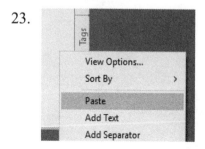 Highlight the Space List – BOMA.

Right click and select Copy.

23. Move the mouse off the tool.

Right click and select **Paste**.

24.

Select the copied tool.

Right click and select **Rename**.

25.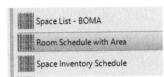

Type **Room Schedule with Area** for the name.

26.

Highlight the new tool.

Right click and select **Properties**.

27.

In the Style field, select the **Room Schedule** style which you just created.

Press **OK**.

28.

Schedule Table			
Room Name	Length	Width	Area
Bathroom	15'-0"	5'-1"	67.72 SF
	11'-9"	5'-11"	80.58 SF
Bedroom	15'-8"	15'-11"	238.44 SF
	11'-4"	12'-8"	143.35 SF
	15'-8"	13'-11"	184.85 SF
Closet	5'-4"	2'-0"	10.82 SF
	4'-9"	3'-1"	15.54 SF
	11'-4"	1'-6"	18.56 SF
	6'-1"	2'-0"	10.32 SF
	3'-1"	1'-2"	4.87 SF
Den	18'-7"	13'-3"	244.74 SF
Dining room	20'-11"	9'-11"	170.74 SF
Garage	20'-4"	20'-4"	408.47 SF
Hallway	3'-1"	18'-0"	49.73 SF
Kitchen	13'-8"	13'-6"	91.89 SF
Laundry room	20'-4"	7'-6"	152.06 SF
Living room	45'-11"	27'-5"	800.54 SF

Activate the Model tab.

Select the Room Schedule with Area tool.

Window around the building.

Press ENTER.

Pan over to an empty space in the drawing and left pick to place the schedule.

Press ENTER.

29.

Activate the **A102 03 Room Plan** layout tab.

30.

Activate the Layout ribbon.

Place a rectangular viewport for the room schedule.

Rectangular

31.

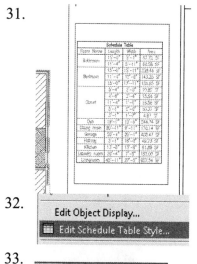

Use Zoom Extents and Zoom Window to center the schedule in the viewport.

32.

Double click inside the viewport to activate model space.
Select the schedule.
Right click and select **Edit Schedule Table Style**.
Select the **Columns** tab.

33.

Highlight the **AREA** column.

Press **Modify**.

34.

Place a check next to **Total**.

Press **OK**.

35.

Select the Layout tab.

Change the Table Title to **Room Schedule**.

36.

ROOM SCHEDULE

ROOM NAME	LENGTH	WIDTH	HEIGHT	AREA

Save as *ex8-23.dwg*.

Layer States

Layer states are used to save configurations of layer properties and their state (ON/OFF/FROZEN/LOCKED). For example, you can set a layer to be a different color depending on the state. Instead of constantly resetting which layers are on or off, you can save your settings in layer states and apply them to layouts.

Layer States are saved in the drawing. To share them across drawings, you need to export them. Each layer state has its own LAS file. To export a layer state, select it in the Layer States Manager and choose Export.

To import a layer state, open the Layer States Manager and click the Import tool. Then select the *.las file you wish to import.

You can save layer states in your drawing template.

Autodesk AutoCAD Architecture 2018 Fundamentals

Exercise 8-24:

Creating Layer States

Drawing Name: ex8-23.dwg
Estimated Time: 15 minutes

This exercise reinforces the following skills:

- ❑ Layer Properties
- ❑ Layer States

1.   Open *ex8-23.dwg.*

We are going to create three layer states: one for rooms, one for floor plans, one for fire rating.

2.   Activate the Room Plan layout tab.

Double click inside the viewport to activate model space.

3.   Activate the Home ribbon.

In the drop-down on the Layers panel, select **New Layer State.**

4.   Type **Rooms** in the Name field.

Type **shows spaces and space tags** in the Description field.

Press **OK**.

5.   Activate the **Floor Plan with Fire Rating**.

Double click inside the viewport to activate model space.

6.   Launch the **Layer Properties Manager**.

7.   Thaw the following layers:

F-Wall-Fire
A-Wall-Iden

8-90

8. Freeze the following layers:

A-Area-Spce
A-Area-Spce-Iden
A-Area-Spce-Patt
A-Anno-Dims
A-Elev-Line
A-Sect-Iden
A-Sect-Line
A-Detl-Iden

9. Launch the Layer States Manager.

10. Select **New**.

11. Type **Fire-Rating Floor Plan** in the Name field.

Type **shows Fire Rating for Walls** in the Description field.

Press **OK**.

12. Close the dialog box.
Activate the A102 02 Floor Plan layout tab.

Double click in the viewport to activate model space.

13. Thaw the following layers:

A-Anno-Dims
A-Detl-Iden
A-Elev-Line
A-Sect-Iden
A-Sect-Line

14. Freeze the following layers:
A-Area-Spce
A-Area-Spce-Iden
A-Area-Spce-Patt
F-Wall-Fire
A-Wall-Iden

15.   Launch the Layer States Manager.

16.   Select **New**.

17.   Type **Floor Plan** in the Name field.

New layer state name:

Floor Plan

Description

Standard Floor Plan

Type shows **Standard Floor Plan** in the Description field.

Press **OK**.

Close the dialog box.

18.   Save the file as *ex8-24.dwg*.

Exercise 8-25:

Applying Layer States to Layouts

Drawing Name: ex8-24.dwg
Estimated Time: 10 minutes

This exercise reinforces the following skills:

- ❑ Layouts
- ❑ Layer States
- ❑ Controlling the display of viewports

1.   Open *ex8-24.dwg*.

2.   Activate the **A102 02 FLOOR PLAN** layout tab.

        Double click inside the viewport to activate model space.

3.   Activate the Home ribbon.

        Set the layer state to **Floor Plan**.

4.   Click outside the viewport to return to paper space.

5. Activate the **Floor Plan with Fire Rating** layout tab.

    Floor Plan with Fire Rating

    Double click inside the viewport to activate model space.

6.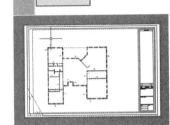

    Set the layer state to **Fire-Rating**.

7. Click outside the viewport to return to paper space.

8. Activate the **A102 03 ROOM PLAN** layout tab.

    A102 03 ROOM PLAN

    Double click inside the viewport to activate model space.

9.

    Set the layer state to **Rooms**.

10. Click outside the viewport to return to paper space.

11. Notice that when you switch layer tabs the different floor plans retain their layer states.

    Save as *8-25.dwg*.

The Sheet Set Manager provides a list of layout tabs available in a drawing or set of drawings and allows you to organize your layouts to be printed.

The Sheet Set Manager allows you to print to a printer as well as to PDF. Autodesk does not license their PDF creation module from Adobe. Instead, they have a partnership agreement with a third-party called Bluebeam. This means that some fonts may not appear correct in your PDF print.

Exercise 8-26:

Creating a PDF Document

Drawing Name: ex8-25.dwg
Estimated Time: 15 minutes

This exercise reinforces the following skills:

- ❏ Printing to PDF
- ❏ Sheet Set Manager

1. Open *ex8-25.dwg*.

2. Type **SSM** to launch the Sheet Set Manager.

3. 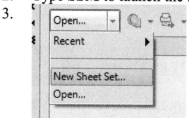 Select the drop-down arrow next to Open and select **New Sheet Set**.

4. Enable **Existing drawings**.

   Press **Next**.

5. Type **Brown Project**.

   Press **Next.**

6. Select the **Browse** button.

7. Locate the ex8-25 drawing file.

   Uncheck all the files.

   Place a check next to the three floor plan layouts:

   Floor Plan with Fire Rating
   A102 02 FLOOR PLAN
   A102 03 ROOM PLAN

8.

Import Options...

Select **Import Options**.

9.

☐ Prefix sheet titles with file name

Uncheck Prefix sheet title with the file name.

Press **OK**.

Press **Next.**

10.

Sheet Set Preview:

Brown Project
1 Floor Plan with Fire Rating
2 A102 02 FLOOR PLAN
3 A102 03 ROOM PLAN

The sheets to be included in the sheet set are listed.

Press **Finish**.

11.

Brown P...

Sheets

Brown Project
1 - Floor Plan with Fire Rating
2 - A102 02 FLOOR PLAN
3 - A102 03 ROOM PLAN

You should see the sheet set list in the Sheet Set Manager.

If you don't, select Open and select the SSM file.

12.

Highlight the SSM file name.

Right click and select **Publish→Publish to PDF**.

13.

File name: Brown Project.pdf

Files of type: PDF Files (*.pdf)

Browse to your classwork folder.

Save the pdf file.

14.

Browse to where the file was saved using Windows File Explorer.

Open the PDF file and review.

Save as *ex8-26.dwg*.

Notes:

QUIZ 4

True or False

1. Columns, braces, and beams are created using Structural Members.

2. When you isolate a Layer User Group, you are freezing all the layers in that group.

3. You can isolate a Layer User Group in ALL Viewports, a Single Viewport, or a Selection Set of Viewports.

4. When placing beams, you can switch the justification in the middle of the command.

5. Standard AutoCAD commands, like COPY, MOVE, and ARRAY cannot be used in ACA.

Multiple Choice

6. Before you can place a beam or column, you must:

   A. Generate a Member Style
   B. Activate the Structural Member Catalog
   C. Select a structural member shape
   D. All of the above

7. Select the character that is OK to use when creating a Structural Member Style Name.

   A. -
   B. ?
   C. =
   D. /

8. Identify the tool shown.

   A. Add Brace
   B. Add Beam
   C. Add Column
   D. Structural Member Catalog

9. Setting a Layer Key

   A. Controls which layer AEC objects will be placed on.
   B. Determines the layer names created.
   C. Sets layer properties.
   D. All of the above.

10. There are two types of stairs used in residential buildings:

    A. INSIDE and OUTSIDE

    B. METAL and WOOD

    C. MAIN and SERVICE

    D. FLOATING and STATIONARY

11. Select the stair type that does not exist from the list below:

    A. Straight-run

    B. M Stairs

    C. L Stairs

    D. U Stairs

12. The first point selected when placing a set of stairs is:

    A. The foot/bottom of the stairs

    B. The head/top of the stairs

    C. The center point of the stairs

    D. Depends on the property settings

ANSWERS:

1) T; 2) F; 3) T; 4) T; 5) F; 6) D; 7) A; 8) A; 9) A; 10) C; 11) B; 12) D